MONOGRAPHS OF THE
SOCIETY FOR RESEARCH IN
CHILD DEVELOPMENT

SERIAL NO. 223, VOL. 56, NO. 1, 1991

HIDDEN SKILLS:
A DYNAMIC SYSTEMS ANALYSIS
OF TREADMILL STEPPING
DURING THE FIRST YEAR

ESTHER THELEN
BEVERLY D. ULRICH

WITH COMMENTARY BY
PETER H. WOLFF

MONOGRAPHS OF THE SOCIETY FOR RESEARCH IN CHILD
DEVELOPMENT, SERIAL NO. 223, VOL. 56, NO. 1, 1991

CONTENTS

ABSTRACT

THELEN, ESTHER, and ULRICH, BEVERLY D. Hidden Skills: A Dynamic Systems Analysis of Treadmill Stepping during the First Year. With Commentary by PETER H. WOLFF. *Monographs of the Society for Research in Child Development*, 1991, **56**(1, Serial No. 223).

When prelocomotor infants are supported on a motorized treadmill, they perform well-coordinated, alternating stepping movements that are kinematically similar to upright bipedal locomotion. This behavior appeared to be a component of independent walking that could not be recognized without the facilitating context of the treadmill. To understand the ontogenetic origins of treadmill stepping and its relation to later locomotion, we conducted a longitudinal study using an experimental strategy explicitly derived from dynamic systems theory. Dynamic systems theory postulates that new forms in behavior emerge from the cooperative interactions of multiple components within a task context. This approach focuses on the transitions, often nonlinear, where one preferred mode of behavior is replaced by a new form. Specific predictions about these transitions help uncover the processes by which development proceeds. Chapters II, III, and IV introduce dynamic principles of pattern formation and their application to development.

In our application of these principles, we tested nine normal infants twice each month beginning from month 1 in a task where the treadmill speed was gradually scaled up and in an additional condition where each leg was driven by the treadmill at a different speed. Kinematic variables were derived from computerized movement analysis equipment and videotaped records. We also collected a number of anthropometric measurements, Bayley motor scores, and a behavioral mood scale for each month.

Several infants stepped on the treadmill in their first month, but in all infants performance showed a rapidly rising slope from month 3 to month

6. Infants also showed corresponding improvement in adjustments to speed and relative coordination between the legs. In dynamic terminology, we found evidence that alternating stepping on the treadmill became an increasingly stable *attractor* during the middle months of the first year. Dynamic predictions that transitions would be characterized by increased variability and sensitivity to perturbation were borne out.

Identifying the transitions enabled us to suggest a *control parameter* or variable moving the system into the stable response to the treadmill. This appeared to be the waning of flexor dominance in the legs during posture and movement that allowed the leg to be stretched back on the treadmill and so elicited the bilaterally alternating response. Further studies are needed to test this hypothesis.

This dynamic analysis confirmed earlier suggestions that skill in general, and locomotion in particular, develops from the confluence of many participating elements and showed how emergent forms may result from changes in nonspecific components. A dynamic approach may be useful for understanding ontogenetic processes in other domains as well.

I. INTRODUCTION

This *Monograph* serves two purposes. First, we seek a better understanding of the ontogeny of walking, a fundamental motor skill. The acquisition of independent upright locomotion has been an important concern in the scientific study of development for over half a century. Learning to walk is a dramatic behavioral milestone whose development illustrates many classic problems: the nature of developmental stages and transitions, the role of maturation versus experience, regressions in development, and the sources of ontogenetic change.

Second, we introduce a new developmental theory and a strategy for examining developmental processes explicitly derived from the contemporary study of dynamic systems—in particular, *synergetics* or *dynamic pattern theory*. Because we believe that this approach has great potential for understanding development not only in the motor domain but in ontogenetic processes in general, we have tried to interweave the theoretical principles with an empirical approach and with a data set that could contribute to our understanding of a real skill. We begin by setting the theoretical context.

WHAT MAKES DEVELOPMENT HAPPEN?

Developmental psychologists study many aspects of the developing infant and child. We ask how children's minds grow and how they come to acquire language and both the commonsense and the formal thought patterns of their cultures. We want to know how children become socialized into the worlds of parents, siblings, peers, and institutions. We try to understand their emotional changes and their increasingly complex perceptual and motor skills. And we seek clues to why children differ and why some children will suffer adverse mental or physical health or troubled social relationships.

What unites a discipline of such diverse concerns is the common focus

1

on processes of change. Whatever the domain, a central question is how infants and children acquire new behaviors. What are the processes that lead to the first words, the ability to do arithmetic, the emotional bonds with the parents, or the skills to do gymnastics or to read? In each case, the behaviors represent novel forms—actions that the child did not do previously. Where do these novel forms come from?

We traditionally believe that development is a product of the interaction of the organism and the environment. Just how the organism and the environment interact to produce new forms is not well specified, however (Oyama, 1985). What is the specific or nonspecific information that the child picks up from the world that causes change? What are the organic requirements for acquiring new and more complex behaviors? What is the relation between growth and development? What are the processes that allow new skills to replace old ones?

This fundamental puzzle has a number of corollary questions that have been of long-standing interest in all developmental domains. First, at one level, development appears to be the accumulation of new, discrete skills: the child walks or does not, reads or does not. This gives development a distinct and real stage-like quality, and this "staginess" is a prominent feature of virtually all descriptive and theoretical accounts. Nonetheless, developing children and their environments must be continuous in time. It is a major theoretical challenge to account for discontinuities in performance arising from processes that are themselves continuous.

The paradoxical nature of developmental stages is further apparent when the transitions to new forms are carefully mapped. What appears as a clear shift at one level of analysis may be much more ambiguous on closer examination. For example, if a child can solve a Piagetian volume conservation task only after training or after the task is modestly simplified, does the child "have" volume conservation? Does the infant who clings to Mom and protests separation in the laboratory but happily stays with the baby-sitter at home "have" separation anxiety? How do we describe the linguistic stages of children who use more complex utterances to describe familiar events than unfamiliar ones? These and many other commonly documented examples of *décalage* or context-specific performance have prompted some theorists (e.g., Brainerd, 1978; Turiel & Davidson, 1986) to discount stages as developmental realities, that is, as qualitative structural changes in the organism, and to suggest instead that these are categories of performance imposed by the observer. The resolution of the stage debate is, of course, inextricably tied to progress on the more fundamental issue, What causes organisms to develop?

Finally, the basic issue of the origin of forms raises questions about the universality of development and the sources of individual differences. Just as we may ask whether developmental stages are a reality when the transitions are so blurred, we may also inquire what the universals of development are given the wide range of individual variability in both process and outcome. Variability in developmental outcome is well documented in any number of areas; it is most dramatically illustrated by our relatively poor predictions of the future well-being of infants and children at risk (Sameroff, 1983). Variability in process is less well understood. This means that children may reach similar outcomes by different pathways. This was the conclusion, for example, in Siegler and Jenkins's (1989) study of how children acquire skill in arithmetic. These authors gave preschool children practice in solving addition problems over several months and individually tracked their strategies for solution. With this practice, most children "discovered" new and more efficient solutions, but the variability of the process was enormous: "The variability in the children's discovery processes was more striking than the commonalities" (p. 106). The microgenesis procedure used by Siegler and Jenkins allowed them to dissect processes of change, but traditional study designs may not uncover either diverging or converging pathways, so the extent of common versus individual developmental trajectories is unknown.

THEORIES OF PROCESSES OF CHANGE

These puzzles and problems—the origins of new form, continuity and discontinuity, and variability—have been long-standing concerns in every domain of interest to developmentalists. Nonetheless, we venture to say that the great bulk of Western developmental science has proceeded to make significant empirical gains documenting new behavioral forms and their precursors, successors, and variants while remaining less concerned, and for the most part atheoretical, about issues of fundamental processes of change.

Jean Piaget stands, of course, as a monumental exception. Piaget's concept of *equilibration,* which he elaborated most thoroughly in his later works (cf. Chapman, 1988; Piaget, 1985), deals directly with the process involved in the acquisition of knowledge. For Piaget, "equilibration" refers to "a process which leads from a state of near equilibrium to a qualitatively different state at equilibrium by way of multiple disequilibria and reequilibrations" (1985, p. 3). Readers will note many strong and recurrent similarities between the dynamic systems approach and Piaget's process of equilibration. Both are derived from thermodynamic principles that recognize that order and complexity in a system must arise in opposition to the continually disorganizing tendency of the universe and that such systems must "suck in" resources from the environment. Both rely on the inherent self-organizing abilities of order-producing systems. Both are essentially dialectic, proposing that new forms can arise only from perturbations that disrupt the stability of the old forms. And both acknowledge the natural ability of such systems to "seek" stable new solutions autonomously, solutions that may be more complex, efficient, or effective than the old forms.

Dynamic systems theory explicitly differs from Piagetian theory, however, in a very fundamental way. In particular, we make no assumptions about mental or other structures; indeed, we eschew the reality of hardwired structures in motor behavior, cognition, or other domains (see Fogel & Thelen, 1987; Thelen & Smith, in preparation). Structural theories, including Piagetian theory, are essentially prescriptive. That is, they assume that the motors pushing the developmental system forward are specific mental structures, representations, schemes, skill levels, programs, or codes that are variously elaborated or enriched by information gathered by the child through experience and that exist independently of their behavioral instantiation.

Our view is much more dynamic. We replace the notion of structures with one of "softly assembled" behavioral *attractors* that can be variously stable and unstable. To call something a "behavioral attractor" makes a probabilistic statement about the preferred performance of the organism under specified conditions. As we shall specify below, we assume that behavior is assembled from component systems that are free to reassemble in different functional ways in different contexts. Thus, the behavior does not "exist" in any privileged form outside its context-specific expression. We believe that the attractor construct, which is first and foremost operationally defined, offers a way to explain the classic puzzles of development and at the same time achieves a better fit to a wide host of empirical data.

The legacy of Piaget is a fully articulated and comprehensive theory of the structure and origins of knowledge. The dynamic systems approach, while congruent with many Piagetian concepts, is in a much less developed state of elaboration. It offers, however, promise of great generality and may be applicable to many domains of development. This generality may be both the weakness and the strength of dynamic systems. The weakness is that the approach itself will have nothing to say about the unique content domain and will not offer a priori predictions about particular behavioral transformations. These can be derived only from empirical study. The strength, on the other hand, is a principled account of how change may occur and the principles that lead to an operational strategy for unpacking developmental process. Because this strategy is not domain specific, it allows for the kind of cross-domain investigations that we shall argue are essential for understanding human ontogeny.

Readers will also note the intellectual heritage of dynamic systems in other organismic approaches to development, especially those influenced by general systems theory (e.g., Bertalanffy, 1968; Laszlo, 1972; Weiss,

1969). A number of prominent theorists have applied contemporary dynamic principles to issues in human development (e.g., Brent, 1978; Kitchener, 1982; Lerner, 1978; Sameroff, 1983; Wolff, 1987). The current work extends and expands on these statements by explicitly developing principles of one branch of dynamic systems, synergetics, to the developmental domain and by showing how these principles can guide empirical research on a classic developmental problem—the onset of independent locomotion. Specifically, we trace the ontogenetic course of one component of locomotion, the dynamic alternation of the legs in a stepping pattern, that is normally "hidden" by other competing processes. We show how locomotion can be explained only as an emergent, not a prescriptive, process, one that depends on the confluence of many asynchronously developing elements.

In Chapter II, we present an overview of dynamic systems, using both commonplace and more exotic examples. In Chapter III, we apply dynamic principles to developing organisms, concluding with a specific translation of synergetic principles into operational guidelines for conducting developmental inquiry. In Chapter IV, we introduce the problem of the developmental origins of locomotion and the particular phenomenon of treadmill-elicited stepping, and we outline our empirical study. Chapters V and VI present the methods and results, respectively, of this study. In Chapter VII, we discuss the implications of our results for understanding locomotion and conclude with a more general evaluation of dynamic systems for other domains.

II. DYNAMIC PATTERN FORMATION IN PHYSICAL, BIOLOGICAL, AND COMPUTATIONAL SYSTEMS

In the last few years, concepts of dynamic systems have captured the imaginations of mathematicians and scientists across a wide number of fields, including physics, astronomy, chemistry, meteorology, and the biological, cognitive, neurological, and social sciences. In their classic form, dynamic systems are systems that change over time; the mathematical foundations for studying time-evolving systems can be traced back to Newton and the French mathematician Poincaré. In their contemporary guises, dynamic systems are also understood to be complex and nonlinear but characterized by relatively simple mathematical formulations. The study of dynamic systems is emerging as a new branch of science not only because such systems display elegant and intriguing mathematical properties but also because they are being used to characterize an increasing number of natural phenomena, from weather systems to physiological processes.[1] In this chapter, we introduce principles of dynamic systems in reference to a few, easily visualized physical systems. We then digress for other illustrations of self-organization in biological and computational systems. Finally, we return to a more formal list of principles for understanding the emergence of behavioral forms in real-time action and in development.

The contemporary study of dynamic systems addresses the most fundamental of cosmological questions—the origins of complexity in the universe. How did the patterns and organization we see in the physical and biological world around us come into being from the random and featureless primeval universe? The same fundamental question is, of course, the basis for devel-

---

[1] For more background, readers are referred to several lively and approachable accounts of this growing interest in dynamic systems and of the accompanying scientific and philosophical debates. The classic introduction is Prigogine and Stengers (1984). Additional popular accounts of mathematical, physical, and biological dynamic systems can be found in Davies (1988) and Gleick (1987). More mathematical general treatments are Glass and Mackey (1988), Grebogi, Ott, and Yorke (1987), Haken (1977, 1983), Kelso, Mandell, and Shlesinger (1988), Shaw (1984), and Yates (1987).

opmental inquiry. What are the origins of new forms in behavior? How do the complex structures and actions of an animal arise from a nearly feature-less egg?

PATTERN FORMATION IN PHYSICAL SYSTEMS

The key to the origins of complexity lies in the abilities of physical matter—under certain thermodynamic conditions—to form itself spontane-ously into patterns whereby the original elements cooperate to produce an organization not contained or apparent in the individual units. It is instruc-tive to think about such self-organization in terms of a few commonplace physical systems: fluid being heated in a pot, water coursing through a pipe, and water dripping from a faucet. Let us consider the first two fluid exam-ples first.

A pot of water contains a very large number of molecules. At room temperature, the molecules bump around randomly, and, as long as the temperature does not change, the overall behavior of the molecules stays the same (although the movement of individual molecules is indeterminate). When the pot of water is placed on the stove with the burner on, the bottom layer of water gets warmer and less dense and tries to rise. If the tempera-ture difference between the bottom layer and the top is not great, the viscos-ity of the fluid prevents any rising. As the bottom layer becomes hotter, however, a critical point is reached where the warmer water begins to rise, and, under certain conditions, this convection will form itself into a stable and ordered pattern of rolls or hexagonal cells. With further warming, this pattern too becomes unstable, and the pot boils with another, seemingly chaotic, pattern. Similarly, imagine the homogeneous flow of water through a pipe. At first, no pattern is evident. As the water pressure is increased, the flow takes on a pattern of neatly ordered layers. With increasing pres-sure, the water undergoes yet another transition to wild vortices and eddies, the hallmark of *turbulent flow.*

Self-organization.—These two unremarkable and everyday phenomena illustrate fundamental dynamic principles. First, there are huge numbers of water molecules in the pot or in the pipe, and each molecule is theoreti-cally free to move about independently and unpredictably. The molecular elements, therefore, are "noisy," and the assembly of water molecules has potentially very many degrees of freedom or ways to combine together. In the specific pot or pipe, however, when heat or pressure is applied, the molecules cease acting as individual elements that bounce around randomly and appear, as if by magic, to cooperate with one another. The degrees of freedom are drastically curtailed: as the elements are "sucked in" to play a particular role in the pattern, they are no longer free to act independently.

Of course, there is no plan or blueprint for convection rolls in a pot of water. How, then, do the molecules "know" what role to play in the ensemble? Pattern formation in fluids is a classic example of self-organization: under certain physical (pot or pipe) and thermodynamic (heat or pressure) conditions, many single elements autonomously act as a unit.

Complex behavior over time.—A second important characteristic of dynamic systems is that the patterns generated in these conditions, while drastically more simple than the behavior of the single elements alone, can show a remarkable ordered complexity as they evolve over time. Homogeneity gives way to rolls and layers, the rolls are formed and dissolved, and layers turn into the whirls of turbulence. Indeed, both the compression of the original degrees of freedom into order and pattern and the dynamic complexity of the resulting pattern can be expressed in precise mathematical formulations. That is, many dynamic systems are characterized by relatively simple equations. However, the behavior of these equations is not simple at all.

Here it is useful to talk about the third everyday plumbing example, the dripping faucet. When the drips come rather slowly from the faucet, the intervals between the drips are constant, so constant, in fact, that dripping water historically has been used as a clock. As in the above examples, under these physical conditions, the water molecules act cooperatively to produce a dynamic pattern. However, as in laminar flow, if we turn the water on harder, the clock-like drips give way to what appears to be a completely random sequencing. But even this apparent randomness has marvelous hidden self-organized order. When the physicist Robert Shaw (1984) plotted many thousands of sequences of drips with the interval between any two drips on one axis and the interval between the next two drips on the other axis (called a "return map"), he found, for certain rates of flow, a structure in the plots like "loopy trails of smoke" (Gleick, 1987, p. 266), not the random scattering that one would expect (Fig. 1d). This means that the patterns of water drips, while fantastically complex, were both cooperative and contingent. Whatever the physical parameters that contribute to the timing of water drips, each drip both contains the history of the previous drips and contributes to the pattern of future drips. As we shall see, this lesson is not lost in development.

Our everyday examples have shown thus far that even seemingly simple physical systems consisting of uniform molecular elements can self-organize into wonderfully complex patterns that change over time in ways that can be mathematically defined. We pause here to introduce some terminology explicitly derived from one branch of the study of dynamic systems, that of synergetics (Haken, 1977, 1983), to describe this self-organizing pattern formation so that we may directly transpose the terminology to developing systems. The theory and methods of synergetics are especially appropriate

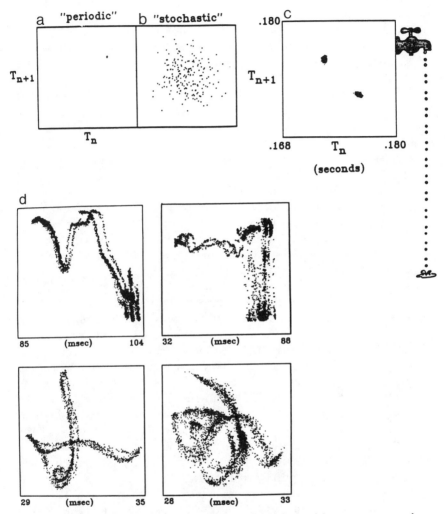

FIG. 1.—Return map of intervals between drips (interval between any two drops plotted against the next interval). *a*, If drips are completely periodic, all points collapse to one place on the space. *b*, If drips were completely random, points would be scattered on the space. *c*, If drips come in pairs, points collapse to two places on the space. *d*, Actual plots of drips at various flow rates. Points are complex, but not random. (From Shaw, 1984; used with permission.)

to biological systems operating over many time scales because, first, they specifically address how ordered behavior is generated from complex systems and, second, the approach has been fully elaborated with a simple human motor activity (Haken, Kelso, & Bunz, 1985; Schöner & Kelso, 1988).

Collective variables and attractor states.—When the multiple degrees of freedom of complex systems are compressed under certain physical and thermodynamic conditions, the resulting system can be described more parsimoniously in terms of one or several variables that express this compression. Haken (1983) has termed these variables the "order parameters" because they appear to order the previously disorganized system. We prefer the alternative term "collective variable" as the dimension of the system that expresses the underlying pattern that emerges from the cooperation of the elements. In the dripping faucet, for example, all the multiple physical forces that determine the drips are captured in the single collective variable of the interval between drips. It is then the behavior of the collective variable over time that is of interest.

Over time, drips could fall from faucets in a huge number of different sequences, from very regular, like drumbeats of different tempo and rhythmic structure, to completely random. In fact, only a small number of these potential patterns is actually seen. At slow flow rates, the drips come one at a time, spaced evenly apart. With a small increase in pressure, they may drip in pairs, and, at yet faster rates, they may show the seemingly random but really structured pattern we described above. In dynamic terms, the particular patterns shown by the system out of the enormous number of possible patterns are said to act as dynamic attractors. In the dripping faucet, the plots of successive drips did not scatter all over the space of the possible combinations of this interval versus the next interval (called the "state space" of this collective variable). If the drips were completely unrelated, the plot would appear as in Figure 1*b*. However, Shaw found that certain regions of the state space were more dense than others. These attractor regions act to absorb, so to speak, other possible sequencing combinations. When drips were regular, all intervals were the same, and the attractor collapsed to a single point, as in Figure 1*a*. When the drips came in pairs, there were two attractor points representing the inter- and intrapair intervals (Fig. 1*c*). The wispy clouds of smoke that emerged from the seemingly random drips represent the now well-known chaotic or strange attractors, whose wonderful characteristics are further described in the references mentioned earlier (Fig. 1*d*).

The attractor concept is central to our dynamic view of behavioral development. It says that complex systems (and we include developing organisms) autonomously prefer certain patterns of behavior strictly as a result of the cooperativeness of the participating elements in a particular context.

Although the dynamic history of the system is important, the attractor states are not encoded or programmed beforehand. Nowhere in the water or the faucet is there a rule or a scheme that produces the elaborate patterns of drip sequences. Attractors are first and foremost *emergent* phenomena.

How do we know it when we see behavioral attractors? Many attractors in physical and biological systems can be (and have been) rigorously characterized. The mathematics for doing so is difficult and requires many more data points than are usually collected in any developmental sequence. We rely here on a somewhat more statistical and qualitative approach, using the relative stability and instability of the collective variables over time and with different contexts as indices of preferred states. Measures of stability and instability rely, in turn, on observing the system at transitions, which we discuss next.

Phase shifts.—In our water examples, the collective variables behave over time with another important characteristic of dynamic systems—nonlinearity, or the ability to change from one pattern to another in a seemingly sudden or discontinuous manner. The transition from homogeneity to layers, rolls, or cells is not gradual; rather, the patterns appear in a sudden phase shift. Put another way, the system shifts between qualitatively different attractor states. This phase shift is engendered, however, by gradual changes in certain physical variables applied to the system, in this case, heat and pressure. The system is sensitive to these influences, termed in synergetics the "control parameters" of the system; the control parameter moves the system through its collective states. "Control parameter" is not the best term because it implies a controller in a machine sense. Rather, the control parameter can be any organic or environmental variable that, when changing, leads to corresponding changes—often, but not always, nonlinear—in the collective behavior of the system. The control parameter does not contain the instructions for the change—again, heat and pressure are entirely nonspecific—but rather moves the system around into new attractor states. We shall propose below that understanding developmental process involves characterizing the control parameters that engender phase shifts in behavioral attractor states.

It is helpful to think of attractors and phase shifts in complex systems as depicted in the potential well diagram of Figure 2. Let us use the boiling pot example again. Imagine that the little ball on the left represents the pot of water when the stove is off and that the steepness of the well represents the amount of energy change the system needs to move it away from its preferred state—in this case, the random movement of water molecules. The walls of the well are very steep because the system is stable: much energy is needed to drive the molecules into another configuration. As, however, heat is added (the control parameter is scaled up), the walls of the well flatten out. Much less energy is needed to change the system. Although

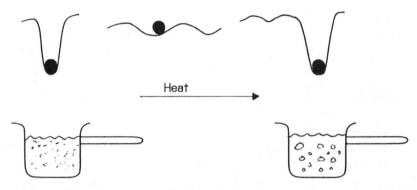

Fɪɢ. 2.—Illustration of the onset of convection rolls in a boiling pot as a ball in a potential well.

the configuration of the collective variable is still the same, the attractor basin is less stable. Finally, a critical point is reached where the system shifts into convection rolls. During this transition, the system is very unstable; the walls are flat, and the ball can roll around anywhere. At this point, the system "seeks" a new stable configuration and rapidly falls into a second, deeper potential well, that of the convection pattern. Again, the steep walls mean that the system must be pushed very hard to change. In a sense, it "prefers" that particular pattern. Commonly, the time taken by transition—the unstable period—is very rapid compared to the time we observe the system in its stable states. Thus, within a range of temperatures the system is stable, but at a critical point the heat disrupts the coherence of the attractor, leaving the elements to discover new attractor states.

Phase shifts in complex systems may result in order from randomness, as in the boiling pot example, or in the replacement of one patterned behavior with another. The most accessible example of the latter is gait change in quadrupeds. When horses, for instance, increase their speed of locomotion, they spontaneously shift from one footfall pattern to another: from a walk, to a trot, to a gallop. There are no intermediate gait styles. Walking is stable within a range of speeds, but at a critical point the animal shifts into a trot, a new pattern. Phase shifts may also involve the generation or loss of multiply stable attractor states or multiple regions of stability on the state space.

Phase shifts are of crucial importance because it is at such transitions that we learn about the nature of dynamic systems. Recall that a coherent pattern damps down the inherent noise and degrees of freedom of the component elements: the components act as a piece. At transitions, however, the noise and variability of the system are revealed. Again, we refer to the ball in the potential well (Fig. 2). When the attractor is stable, the ball is in one place; at a transition, the ball can move all around. This means that there will be more variability around the mean values of the collective variable. Similarly, if we give the ball a little nudge when it is in the deep well, it will rapidly seek the minima. In the shallow well, the same nudge will really move the ball around, and it will not find the minimum potential as rapidly. Synergetic theory predicts, therefore, that systems at transition will be both more variable and more responsive to perturbation than those in stable attractors. We shall see that this prediction is a powerful tool for unpacking developmental process.

The thermodynamics of self-organization.—The emergence of pattern from disorder flies in the face of what we all know about the universe from the second law of thermodynamics. By this law, the direction of change is from order to disorder. That is, the disorder in the universe, its entropy, must be increasing. Heat dissipates: ice melts in warm water because the heat flows from the water to the ice. The reverse is never true; heat never flows from ice to make the water warmer. Energy goes from a source to a sink in a one-way direction. Yet the coherent patterns in the pot and the pipe reverse this arrow of time. Order is increasing, not decreasing. How can this be?

The second law of thermodynamics is absolutely true in a closed system, that is, a system where energy is neither added nor removed. This may be conceived as our pot of water before the burner is turned on. Any initial heat in the water will dissipate to the atmosphere until the water is at equilibrium with the air. Nothing will happen in the pot. However, the addition of heat energy creates a hot spot, a localized source of continuing energy. This is necessary for pattern formation: self-organization can occur only when energy of some form is taken from one source and pumped into the system.

All biological systems, of course, concentrate energy, taking it as food from one source in the universe and using it to maintain order in the face of the second law. Biological and many other systems are in thermodynamic nonequilibrium—they suck up and store energy from their surrounds, an essential condition for the processes we describe.

A note on self-organization in faucets and humans.—We began with the example of physical systems that create patterns from the cooperation of many simple elements because these systems are both commonplace and transparent; we have no need to seek "a ghost in the machine" to explain the origins of drops in a faucet. One might object, however, on the legitimate grounds that developing humans are not faucets and pipes and that, while these physical analogies are all well and good, human behavior and development cannot be reduced to physical principles (see, e.g., Hofsten, 1989). Here, therefore, it is important to clarify the theoretical status of these physical examples vis-à-vis real, developing babies and children, with real genes and real brains. Do we need concepts of self-organization when indeed both genes and brains can encode instructions for development and behavior?

First, as we argued in Chapter I, structural or maturational/genetic accounts of development have not provided a compelling explanation of process, nor have they successfully dealt with the data. Dynamic systems may both be more theoretically sound and explain more data. Second, the physical illustrations are meant as heuristics to present principles of pattern formation, not to suggest that the components or the mechanisms are the same in faucets as in babies. Third, self-organization is ubiquitous in nature in biological as well as physical systems, as we describe in the next section. Finally, processes of self-organization are not mystical or magical and indeed can be elegantly simulated by mathematical and computational means. In the final section, we review a few examples of such computational systems.

PATTERN FORMATION IN BIOLOGICAL SYSTEMS

The remarkable advances in plant and animal genetics in the last several decades, culminating in feats of genetic engineering and the technical tour de force of the human genome project, may leave developmentalists with the feeling that their jobs are redundant. Once the geneticists have mapped every gene on each human chromosome, we may imagine, the fate of the organism will be known and determined. But this is not so. Living organisms produce patterned structure and behavior dynamically, that is, as an emergent process (Oyama, 1985). Development and, ultimately, behavior are as much a product of contingent, nonspecific influences and events as they are of specific genetic transductions to the cells or specific neural impulses to the limbs. These points are made more specifically in the following examples.

Pattern formation in the slime mold.—Slime molds are nearly microscopic plants that live in the soil. They have been extensively studied because their elaborate life cycles, consisting of several morphogenetic forms, embody many important characteristics of embryonic development, but on a more simple and accessible scale. Slime molds first emerge as single-cell amoebas, which feed on soil bacteria. When their food sources become scarce, some of the cells secrete a chemical attractant that leads to an aggregation of the amoebas into a streaming mass. Soon several tip-shaped elevations appear on the mass; these then guide the cell movements and lead to a standing, slug-shaped structure. Just below the tips on this standing structure, cells differentiate into more rigid stalks that then grow up into the air forming

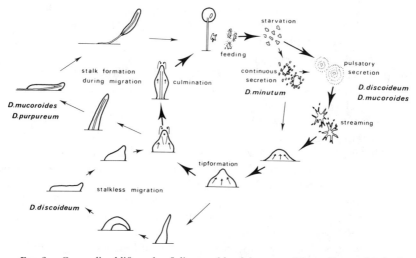

Fig. 3.—Generalized life cycle of slime molds of the genus *Dictyostelium* and *Polysphon-dylium* showing aggregation, migration, tip, and stalk formation. (From Schaap, 1986; used with permission.)

a fruiting body in which the cells turn into spores, and the cycle is repeated. Different species develop different forms of fruiting bodies (Fig. 3).

How do the single cells come together and then differentiate into several functional roles in the aggregation? The process is initiated when some cells respond to starvation by secreting c-AMP, the chemical attractant. In some cells, the production of c-AMP oscillates in an autonomous way. Neighboring cells detect these pulses and in response secrete pulses of their own at the same time that they stream toward the pulse of chemicals. In this way, the signal is passed through the cells, which move toward the center of oscillation, which becomes the tip of the mass. Aggregation, tip formation, and movement, then, appear to be the direct consequence of the chemical attraction of the cells toward an autonomous oscillator. Complex and ordered patterns cascade from the simple ability of some cells to begin to secrete c-AMP in waves (Schaap, 1986).

The differentiation of cells into stalks and spores is also a contingent process. When the cells aggregate, they autonomously sort themselves on the basis of certain nongenetic characteristics, especially the length of starvation, the cell phase at the onset of starvation, and their current metabolic phase. This initial sorting is based on environmentally produced metabolic differences, probably mediated through differences in the adhesion qualities of the cell surfaces. The sorted cells use additional position-dependent cues, probably concentration gradients of simple molecules, to determine the fate of the cells as stalks or spores (Schaap, 1986).

Slime mold morphogenesis illustrates in particularly clear fashion how cascading interactions between the organism and environmental conditions can produce emergent pattern. All that need be encoded in the genetic material is a metabolic system that responds to starvation in certain ways. The life-cycle changes are not programmed in but fall out in a dynamic and contingent way from these initial conditions. Indeed, embryological development in general, including that of higher vertebrates, appears to take advantage of the organizing capabilities of rather simple physical parameters for inducing pattern. These include chemical gradients in the cells, the directional effects of gravity on the egg and embryo, the physical tensions induced by membranes in adjacent cells, changes in viscosity produced by aggregating calcium ions, and so on. The genome does not need to encode all the details of morphogenesis because the system has capitalized on the organized fields of nongenetic organic and environmental parameters. Indeed, pattern formation in embryogenesis has been elegantly modeled using such nonspecific parameters as gradients, polarity, surface tension, packing geometry of cells, and cell migration dynamics (see, e.g., Goel, Doggenweiler, & Thompson, 1986; Goodwin & Trainor, 1985; Meakin, 1986; Mittenthal, 1981; Odell, Oster, Alberch, & Burnside, 1981; Sekimura & Kobuchi, 1986; Soll, 1979; Tapaswi & Saha, 1986; Wilt, 1987).

Insect nest building.—We move now from our single-cell example of emergent pattern to self-organization on a macroscopic, behavioral level, in particular, the nest building of social insects as discussed by Kugler and Turvey (1987). African termites build large, complex nests that consist of many columns and arches covered by a dome. How do these simple insects encode such complex architectural plans? How do they "know" to build in the correct spatial and sequential patterns?

Initially, insects move around the building sites and deposit building material randomly. The deposited materials emit a volatile odor or pheromone, which attracts other termites (Fig. 4). In the presence of the pheromone, the termites behave according to two simple rules: move toward the strongest smell, and deposit material where the smell is strongest. Although the initial deposition is random, soon several deposits act as basins of attraction because of their concentrated odor, and deposited material builds up (Fig. 4b). The diffusion patterns of the pheromone through the nest-building site build up gradients in three-dimensional space that differentially attract insects to deposit at certain sites (Fig. 4c). New deposit sites will be established at the intersections of established sites, where concentrations are high. At the same time, as the construction progresses, the gradient fields become defined and constrained by the sites already built (Fig. 4d, e). In this dynamic way, the construction of pillars, arches, and domes can be understood (and has been modeled) solely as a function of the time-dependent flow of odorant from the deposit sites interacting with the gradient-dependent behavior of depositing insects. No other "knowledge" is necessary for these structures to emerge.

Slime molds and termite nests illustrate features of biological self-organization that will recur in our other examples. These examples span various levels of organization and time scales. It is not necessary for the amoebas or the termites to have a genetic or mental code specifying the morphology of either the colony or the nest because the organisms can take advantage of the organizing properties of more simple biological and physical systems. In the slime mold, the pulsating c-AMP acts as the control parameter that shifts the organisms into a qualitatively new form. In the termites, the simple, scalar properties of gradient fields, in concert with the termites' activity, serve as the nodes for the construction of elaborate nests. Pattern emerges through the cooperation of many simple elements within understandable thermodynamic conditions.

PATTERN FORMATION IN COMPUTATIONAL SYSTEMS

The examples of the slime mold and the termites illustrate both simplicity out of complexity (many elements cooperate to produce a more ordered system) and complexity out of simplicity (the resulting system is structurally or behaviorally complex). These dynamic principles are also well illustrated by two examples of dynamic mathematical models of complex living systems. The first, the logistic equation, was first used to describe ecological population changes, and the second, an example of a neural network, has been used to model a number of human perceptual, cognitive, and motor activities. In each case, complexity and order are generated autonomously.

The logistic equation.—One of the more remarkable illustrations of dynamic nonlinearity is a seemingly simple equation that has been classically used to describe cycles of population change. Ecologists have long been concerned with the natural fluctuations in the numbers of organisms in any ecosystem. If you have, for instance, x number of fish in a pond this year, how many fish will you have next year, and in following years? If population growth were unrestrained, x for next year would be a function of x for this year that depended only on the reproductive rate r of fish, $x_{(next)} = rx$. Growth would increase linearly. Clearly, however, unlimited growth is impossible owing to limited food supply, predation, disease, overcrowding, and so on. In fact, next year's population might increase dramatically if the original population were small and food abundant, but it would decline if the increased population resulted in severe competition. An equation describing population fluctuation needs to include terms that represent not

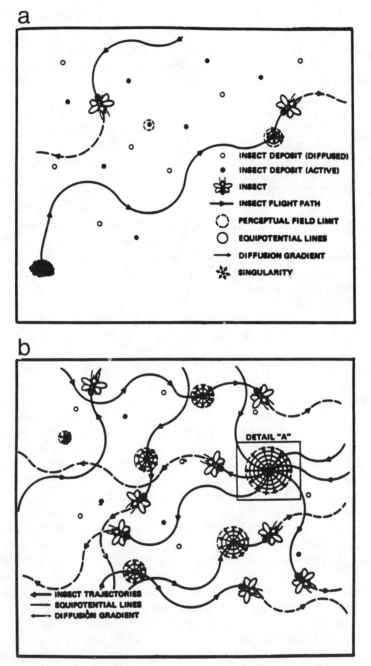

FIG. 4.—Termite nest building as a self-organizing process. *a*, Insect random flight paths and initial deposition sites. *b*, Concentrated deposition along diffusion gradients builds up sites. *c*, Three-dimensional diffusion gradients of pheromone; building a tower. *d*, *e*, Building an arch by depositions at the intersection of three-dimensional deposition sites. (From Kugler & Turvey, 1987; used with permission.)

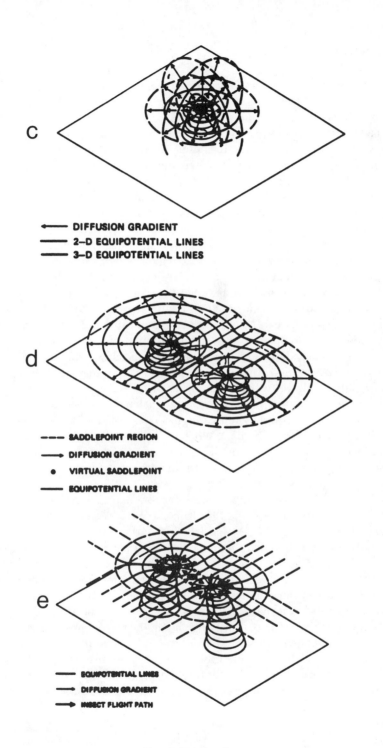

c

DIFFUSION GRADIENT
2–D EQUIPOTENTIAL LINES
3–D EQUIPOTENTIAL LINES

d

SADDLEPOINT REGION
DIFFUSION GRADIENT
VIRTUAL SADDLEPOINT
EQUIPOTENTIAL LINES

e

EQUIPOTENTIAL LINES
DIFFUSION GRADIENT
INSECT FLIGHT PATH

FIG. 4 (*Continued*)

only the reproductive rate but also the death rate from various causes. The simplest version of this is $x_{(next)} = rx(1 - x)$, where the term $(1 - x)$ keeps the population from continually increasing.

The addition of the term $(1 - x)$ gives the so-called logistic equation remarkable dynamic properties, including qualitative phase shifts as a function of changes in a control parameter and the generation of enormous complexity. In this case, the control parameter is r, the growth rate. Using a specific r, the equation is solved iteratively, that is, the x from one year is plugged in to solve for the next year's population level, and so on.

Figure 5 plots the final population level as a function of differing growth rates. At low growth rates, the population becomes extinct. As the growth rate slowly increases, there is a region where the population increases steadily, in a near linear manner. However, as the growth rate passes 3, the system begins to oscillate between two population levels, rising and falling in successive years. Turning up r a little more leads to another bifurcation: cycles every 4 years, then every 8, 16, and so on. Finally, at higher growth rates, the fluctuations appear not to settle into any predictable sequence but to shift wildly and seemingly unpredictably from year to year. However, as r is continually cranked up, there are again regions of periodicity. The system behaves with deterministic chaos—complex, nonlinear behavior that looks random but is not. Chaotic behavior has, of course, been identified in many natural systems ranging from weather patterns to enzyme dynamics and heart rhythms. What is noteworthy in the current example is the generation of this time-dependent complexity from the iteration of a seemingly simple and transparent function.[2]

A neural network model of the acquisition of a skill.—For a second example of computational self-organization, we refer to a neural network model of hand-eye coordination developed by Kuperstein (1988; Grossberg & Kuperstein, 1986). This model is especially interesting for developmentalists because the process by which the model learns to position an arm to reach an object in space is similar to the Piagetian process of circular reactions, where infants develop a correlation between sensation and movement in actions that are repeated many times. The neural network is a system of equations operating on matrices of numbers to represent abstract neural units that initially contain no information about the target's position in space or about the simulated mechanical arm-head-eye system that simulates the reach. According to Kuperstein (1988), learning this sensorimotor skill requires that "representations of posture emerge out of the correlation between posture sensation and target sensation" (p. 1308). In the model, this

---

[2] Van Geert (1991) has recently published a dynamic model of cognitive and language development using logistic growth curves. His mathematical simulations share many assumptions with the present approach.

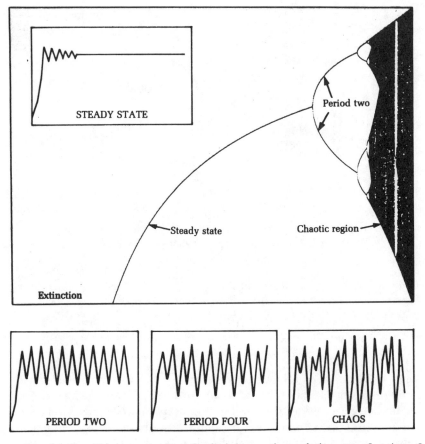

STEADY STATE

Period two

Steady state

Chaotic region

Extinction

PERIOD TWO

PERIOD FOUR

CHAOS

Fig. 5.—Graphic representation of a logistic equation: solutions as a function of changing r, the reproductive rate (from Gleick, 1987; used with permission). R is plotted on the X-axis and the population, x, on the Y-axis. At low r, the population becomes extinct; at moderate values of r, the population increases linearly; at still greater values, the population cycles every 2 years (period 2), leading to chaotic fluctuations.

occurs in a two-stage circular reaction. In the first stage, the system generates random motor activity that allows the simulated arm to explore a large range of arm positions. With each position with the object in hand, sensory information about the target is fed back to a sensory map by means of gating factors (weights) that produce motor signals from the actual positions. The network computes the difference between the self-generated and the target motor signals and adjusts the weights so that these differences are minimized. These weights, which are calculated for all possible hand postures, make the system self-consistent. Thus, in the second phase, when the system

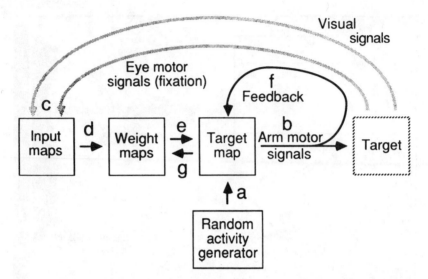

FIG. 6.—Kuperstein's model of a Piagetian circular reaction in a computational simulation. Self-produced motor signals that manipulate an object target are correlated with target sensation signals. The sequence of steps for training is a, b, c, d, (e + f), g. Correlated learning is done in step g. After correlation is achieved, target sensation signals alone can evoke the associated motor signals to manipulate the target accurately. The sequence of steps for performance is c, d, e, b. (From Kuperstein, 1988; used with permission.)

"senses" the position of the target in space, this sensory information is gated through the correlated weights and calls forth the appropriate motor signals for that target position to move the arm and hand (see Fig. 6). This is a self-organizing process because the solutions are not in the system initially; there is no a priori knowledge of the object or of "correct" motor signals. No explicit teacher or error correction is involved. The system "settles" on good performance, given the correct initial architecture of the network and repeated iterations of exploratory performance.

Network models like those of Kuperstein and other so-called distributed parallel models are types of dynamic systems. Whatever these models' degree of neural realism, their message for developmentalists is the generation of emergent properties from elements that themselves do not contain those properties without special vitalist assumptions. In the next chapter, we suggest that this is the hallmark of developing organisms.

III. A DYNAMIC SYSTEMS APPROACH TO DEVELOPMENT

In the previous chapter, we introduced dynamic phenomena in seemingly simple physical systems and used these examples to introduce concepts and terminology appropriate to studying change over time in complex systems. We then attempted to demystify notions of emergent pattern formation by referring to several biological and computational systems. We now return to the synergetic principles of pattern formation in complex systems to outline an approach to development.

A NOTE ON TIME SCALES

The physical and biological examples given earlier illustrated emergent phenomena on several different time scales: seconds and minutes in the pipe and pot, hours in the dripping faucet, and days and weeks in the molds and termites. It is important to emphasize here that the synergetic approach that we introduce in this *Monograph* is meant to span the time scales of both behavior (stepping on the treadmill) and development (the emergence of stepping during infancy). Here we make the strong assumption (following Kugler, Kelso, & Turvey, 1982) that the same processes engendering change over time govern both the second-by-second emergence of action in its accustomed and stable forms and the development of those forms over weeks, months, and years. We will illustrate these dual, but related, time scales first in an everyday example of infant crying and later, in Chapter IV, in our treatment of treadmill stepping.

DYNAMIC PRINCIPLES OF BEHAVIOR AND ACTION

Compression of the degrees of freedom.—Behaving and developing organisms are high-dimensional systems of many anatomical, physiological, and neurological components. Behavior arises from the self-organizing cooperation of these components and subsystems within some environmental and task context. Behavior is thus a compression of the many degrees of freedom potentially available from the combination of the components. A collective variable (or variables) can be identified that expresses this compression. The behavior of the collective variable may itself be complex.

To give one example, when infants cry, the behavior depends on the particular anatomical configurations of the lungs, air passages, and oral cavity, the emotional or motivational state of the infant, the innervation of the muscles, and the directed flow of air through the anatomical structures. Each of these subsystems, in turn, has many elements that would be potentially free to combine with the others in a variety of ways. Yet crying has a particular well-defined acoustic configuration that can be described in much simpler terms than the behavior of all the participating elements, for example, as frequency spectra or amplitude modulations. Although such descriptions are low dimensional compared to the complexity of the contributing elements, they still show crying as complex and dynamic, with elements of rhythmicity, for example (Fogel & Thelen, 1987).

Self-organization.—Self-organization means that behavior emerges strictly as a cooperative function of the subsystems within particular environmental and task contexts. There is no single element that contains the prior instructions for the behavioral performance. The task and the context recruit and assemble the cooperative system, but the essence of the behavior lies in neither the organism nor the environment alone. Thus, neither has logical priority in explaining behavior or its changes.

In crying, for example, we would no more assign the causes of the behavior to a phylogenetic inheritance that includes crying, or to a crying "schema" in the brain, than we would to the safety pin pricking the baby. None of these instantiate crying. Rather, the behavior emerges as the confluence of organic status, motivational propensities, and circumstances. This view contrasts with theories that propose that behavior originates from genetic or mental structures that the organism "has" somewhere in material form. Likewise, in development, the end state of the organism is not materially encoded in the initial state; rather, it arises as the contingent and dynamic organization of form.

Attractors and stability.—Behavior and its ontogenetic course are thus not hard wired or predetermined. However, some ensembles are more stable than others; the cooperative system seems to "prefer" certain configurations and trajectories. In dynamic terminology, behaving and developing systems may settle into certain attractor states, or preferred behavioral outputs, that are determined by the morphology of the system, its particular energetic or motivational states, and the task and environmental contexts. For example, crying is a highly stable response given the anatomical and physiological characteristics of infants when they are stuck with a pin. As long as the pin remains open, all behavioral possibilities are likely (although not obliged) to converge on crying of a particular configuration, which by its rhythmicity, pitch, and amplitude is recognized by parents as a cry of pain. The stability of this attractor state could be empirically and statistically determined (although we are not recommending these experiments!) through the likelihood of this response given the similar context and through perturbation experiments. Efforts to calm the infant will be to no avail if the pin is open. Indeed, the attractor may be so strong that the behavior may persist for a time even after the pin is discovered and removed.

Thus, when the system is assembled in different task contexts, or when the components have changed through growth or differentiation, the stability of the attractor states may change. That is, some behavior becomes more likely to be performed, more tightly constrained, more skilled, and less subject to perturbation. Other patterns, however, may become less reliable and more easily disrupted. Since it is the unique ensemble of the parts in response to a task environment that determines the output behavior, the system is exquisitely sensitive to the states of those components. Thus, under similar conditions, systems can be very stable, but they are also capable of great task flexibility. Again, with crying, the same pin prick that evokes a strong and persistent crying response in an infant who is awake may barely disturb an infant in deep sleep. The change of the state component changes the assembly of the behavior. Over developmental time, the crying attractor stability will also change. Older children and adults would likely respond to a pin prick with an entirely different configuration: a verbal "ouch" or an effort to remove the pin. As we detail below, operationalizing a dynamic approach to development depends on measures of relative stability.

Are attractors just another name for the traditional developmental "stage"? The answer to this is both yes and no. If stages are conceived as a statistical propensity for the developing child to exhibit preferred configurations of behavior under certain circumstances, then yes—the attractor may describe age-related phenomena. If, however, a stage is something reified in structure or thought of as existing anywhere outside the real-time assembly of the behavior, then the attractor notion, while capturing "staginess," is not isomorphic with a stage. Recall that the essence of both Piagetian and neo-Piagetian theory is that some underlying ability defines and directs performance; thus, variations are described as deviations from some modal and appropriate level. The attractor concept makes no assumptions about typical performance but focuses only on relative stability and change. Stages are referenced to an ideal; attractors are referenced to their dynamic status.

Phase shifts.—Developmental change may be linear and gradual, such as the usual growth increments in body weight or size. But development very frequently shows discontinuities, where new abilities emerge from precursors that do not apparently contain those forms. A dynamic approach conceptualizes the discontinuities in development as phase shifts. As in other dynamic systems, when one or more components change beyond certain threshold levels, the entire ensemble may move into a qualitatively new mode of behavior. A phase shift implies a transition between two stable modes, where the intermediate states are more unstable and transitory. It is important to note, however, that whether a particular behavioral change is envisioned as gradual or abrupt depends on the time scale of the observation. The shift to independent locomotion, for example, is rapid when viewed in the scale of the life span or even childhood but more gradual if the infant is observed in attempts to walk day by day or hour by hour.

Also as in other dynamic systems, phase shifts in development are engendered by only one or a few of the components of the system, termed the "control parameters." Again, these do not control the system in the sense of a command or a prescription. The system's integrity is sensitive to the control parameter elements or conditions such that, when they reach certain critical values, the internal cohesion of the system may be threatened. Recall that there is inherent noise associated with all the participating components in a complex system but that this noise is damped or compressed when the system is configured in a stable attractor. At critical values of the control parameter, however, this noise becomes amplified, and the system is free to explore other patterns and seek new stable states: a phase shift. Such transitions can be detected by the increased behavioral variability and perturbability of the system. That is, at the time of transition, the system should be in more regions of its possible state space and should show less damped reactions to internal and external perturbations.

Imagine, again, our young infant perfectly awake. We could envision scalar increases in applying pressure on the skin at a point, ranging from a brush to a prick. At low pressure, the stimulus may be ignored. At somewhat higher pressures, the stimulus might be noticed but without an ensuing reaction. However, there will be a critical pressure that will elicit full-blown crying, a behavioral shift. This critical point, in turn, may be much higher in infants in deep sleep. On a developmental time scale, we would be able to identify some age where self-control and social pressure would raise the threshold of crying so high that it is inhibited with, say, an inoculation but not with more serious injury. In each of these examples, there is some point at which stability of the noncrying state is threatened by the noxious stimulus and a new attractor emerges in rapid fashion. Note here that both initial state (awake or asleep) and the pressure of the stimulus may act as control parameters—at the same stimulus level, the state will shift the response; at the same state level, the stimulus will shift the response. Crying is contained neither in the organism nor in the environment alone but only in their confluence.

Development as the stabilizing and destabilizing of attractors.—At the macro level, then, development can be envisioned as the recurrent stabilizing and destabilizing of behavioral attractors, defined in terms of appropriate collective variables. The attractor basins act like discontinuities in the potential landscape, but they are the product of underlying processes that are continuous in nature. Some behavioral configurations lose stability as the underlying components change; other new forms become established. The stable and reliable patterns of crying seen in newborns and young infants dissolve and are replaced by new patterns. Early crying seems so stable and repeatable both within and among infants that it is easy to imagine that it is "hard wired" in the nervous system. However, crying, as are all other behaviors, is so easily shifted by context and so sensitive to the state of the organism that this "hard-wired" explanation is insufficient (see also Fogel & Thelen, 1987). Rather, it is more useful to imagine a "soft assembly" of components, free to assemble in any number of ways, but indeed "preferring" only limited places on a state space and assuming configurations of varying stabilities and instabilities.

Development as a dynamic, multilevel, nonstationary process.—At any point in time, behavior is assembled from component structures and processes within a task context. Over time, however, the components themselves change and reorganize. This can be imagined as layers of parallel dynamic processes, each with its own trajectory (Fig. 7). Some components may increase in a more-or-less linear fashion. Some elements may be characterized by rapid changes followed by plateaus, while others may show decline. In crying, for example, control of the respiratory apparatus may have a very different ontogenetic course than control of crying as social communication (Fogel & Thelen, 1987). Because the behavioral output is assembled by the demands of the context and task as the most stable ensemble of the available components, developmental change need not be paced by the transformation of a single precursor of overarching mental or neural structure. Rather, it may be a small change in any of the system-sensitive components that acts to generate new forms of behavior (or eliminate old forms). These system-sensitive components may well be nonobvious precursors or supportive contexts.

Since the processes and structures used in assembling the behavior are themselves not stationary but dynamically changing, it is useful to think of the whole developing system as migrating through its abstract state space, where different subsystems, including social and environmental influences, may act as control parameters at different times. Because components develop in an asynchronous manner, and because all of them contribute to the performance of any target behavior in an appropriate context, it is possible that certain elements in that target behavior will be potentially available but normally unexpressed in the usual context of the behavior. In

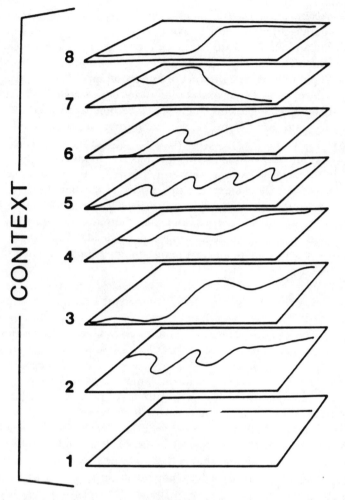

Fig. 7.—Depicting development as parallel developing subsystems, each with its own trajectory. Time is abstractly represented on the X-axis; an abstract "quantity" of a contributing subsystem is represented on the Y-axis. Behavior at any point in time is assembled from these subsystems within a task and environmental context.

general, the slowest-maturing component must reach a critical threshold value for the system to be functionally assembled. The slowest-maturing component in developing systems has been termed the "rate-limiting variable" by embryologists (Soll, 1979). Note that the rate-limiting variable can be, and often is, the control parameter, that is, the crucial scalar variable that engenders new behavioral modes. But the control parameter can be more than just the slowest-maturing component. For example, in many cases it is not the organic substrate that limits performance but rather some accumulation of knowledge or experience. In this case, the control parameter is, at least partially, environmentally determined. Here component elements may potentially support a new behavioral form, but the appropriate task or environmental context does not assemble the available structures and processes. Finally, we must acknowledge the ability of dynamically assembled organisms to meet specific task challenges in a variety of fluid and often creative ways. Since the components are in principle free to reassemble in nonprescribed ways, task goals may be met by novel means even when normally used elements are unavailable.

There are many examples in the developmental literature of both the heterochronicity of developmental components and the fluidity of behavioral performance. It is quite common for elements of a target skill to be detected before the functional performance of that skill, usually when the task context, demands, or instructions are altered, often in a quite subtle way (see Fogel & Thelen, 1987; Thelen, 1989; Turiel & Davidson, 1986). Similarly, aspects of performance may be lost with equally subtle manipulations. Our dynamic approach sees this pervasive phenomenon of décalage not as noise or aberrations from the inevitable sequence of encompassing stages. Rather, such nonlinearities are the inevitable consequences of a process that is multidetermined and asynchronous and in which output is functionally assembled rather than schema driven. When experimenters simplify the task or provide environmental cues or supports, they "fool" the child into precocious performance by substituting for elements that are not otherwise available. We will maintain in this *Monograph* that treadmill stepping— and the development of walking—can be understood only by such dynamic processes.

TRANSLATING DYNAMIC PRINCIPLES INTO EMPIRICAL STRATEGIES

Dynamic systems analyses focus, therefore, on the processes that drive development forward. The concern is not so much with the end state as with the paths by which the organism moves through the life cycle. Indeed, dynamics means there is no end state (although there may be very stable attractors), just relative stability and change. The most important assumption is that development is always multiply determined and that no component or subsystem has ontological priority. The task, the environment, and the organism are all logically equivalent as potential elements for change. The essential nonlinearity of developing systems is reflected both in the capacity for the system to self-organize and in changes of state from the loss of stability.

So far, we have outlined a set of principles for development that are derived from pattern formation in complex, nonlinear systems. These ideas have surfaced in developmental theory before. While such concepts as "equilibration" or "dialectics" have a satisfying resonance, they have been neither operationally defined nor empirically tested (Thelen, 1989). What we show in this *Monograph* is one way to translate these abstractions into a useful empirical strategy with the potential for unpacking developmental process. We emphasize again that dynamic principles alone will tell us nothing about particular developmental phenomena. Dynamic systems conceptualizations cannot predict the specifics of the behavior of the collective variable, its transitions, or the nature of the control variables. These must be revealed by the usual painstaking descriptive and experimental studies. So, although abstractions can never substitute for data, they can provide a theoretically informed method of search.

The overall strategy for operationalizing dynamic principles for studying human development is to identify the collective variable and its attractor states as they change over time. The point is to discover phase shifts and to use the instabilities of the system at phase shifts to identify and manipulate experimentally the control parameters that engender the shifts.

A crucial assumption in a dynamic strategy is that the individual and his or her behavioral changes over time are the fundamental units of study. The primary data for a dynamic approach is the developmental trajectory of the collective variable of interest. It is common in developmental studies to collect cross-sectional data comparing performance on defined tasks at various ages. While this may be an essential first step toward mapping the boundaries of the behavior of interest, individual data, collected at frequent intervals in a longitudinal design, are essential for several reasons. First, analyzing the initial and end points of group data will mask the important potential variabilities in the pathways toward the goal performance. Recall that, in dynamics, stable attractors appear to "pull in" trajectories. If human development is dynamically organized, we would expect that stable performance in, say, language or walking will result despite a variety of initial conditions and via a variety of trajectories through the performance space. Thus, individual variability is not noise to be written away as error variance, as it is in our typical ANOVA models of inferential statistics. Rather, this variability is the data itself since it informs us of the relative stability and instability of the collective variable.

OUTLINE OF A DYNAMIC STRATEGY

More specifically, a dynamic strategy includes the following steps.

1. *Identify the collective variable of interest.*—Recall that, in dynamic systems, one or only a few variables can be identified that capture the compres-

sion of the degrees of freedom of a multidimensional system. It is the behavior of this collective variable over time—its stability and change—that is of primary interest. In simple physical systems, identifying the collective variable may be relatively straightforward: time intervals between water drips, for example. In developing systems, this is a difficult and nontrivial problem because the systems themselves are nonstationary. That is, because of the underlying nonlinearities in the components, a variable may index something different at one age than at earlier or later ages. Developmentalists have long recognized the difficulty of choosing age-appropriate tasks for comparing performance. In the present study, the collective variable is the phasing of alternating steps on the treadmill, which we believe is a reasonably stationary collective variable to describe interlimb coordination and, on the basis of our pilot studies, one for which we expected to see developmental changes. However, in less clear cases, preliminary observations and experiments may be necessary to determine what best captures the behavior of interest and its development.

2. *Characterize the behavioral attractor states.*—This step involves mapping the preferred states of the collective variable to ask what stable behavioral regimes characterize the organism at particular points in time. Here is where cross-sectional studies are important. Before beginning to collect longitudinal trajectories, it may be useful to know the relative stability of the collective variable at one or more target ages. Stability may be assessed either between individuals (how likely is this behavior within a group of infants or children), within an individual (how likely is this individual to repeat this behavior between test trials or on different occasions), or by manipulations that perturb the collective state. Behavior that is stable between and within individuals and to which the system returns when perturbed acts like an attractor. In Chapter IV, we report on our background studies mapping the stability of the treadmill attractor state.

3. *Describe the dynamic trajectory of the collective variable.*—Dynamic process accounts require frequent time-based data points that track the state of the collective variable. This requires longitudinal studies with frequent sampling. The time span for the data collection depends, of course, on the time scale of the ontogenetic change of interest, which may be determined by cross-sectional studies. When behavior is changing rapidly, as in infancy, the longitudinal study may last for only a few weeks or months. The primary data, then, are individual trajectories of the collective variable over time. A significant portion of the data we report in this study is concerned with mapping the developmental trajectory of treadmill stepping.

4. *Identify points of transition (transitions are characterized by loss of stability).*—The goal of longitudinal studies is to identify the time or times when the collective variable undergoes a transition or phase shift. The phase shift signifies the emergence of a new behavioral form. Dynamic theory predicts

that, at transitions, the system is unstable: the coherence of the components in the particular context is lost. This loss of stability can be detected as enhanced fluctuations around the preferred attractor state. Operationally, this would be reflected in larger variability measures within each individual's performance. In addition, unstable systems are more vulnerable to perturbation. Thus, at transitions, experimental perturbation would result either in a longer recovery time for the system to settle back to the preferred state or, with a sufficient nudge, in pushing the system into a new, stable attractor.

A dynamic approach, therefore, replaces descriptions of ideal stages with measures of relative stability and change. The approach elevates variability from the status of noise to that of essential data. As we show in this *Monograph*, individual differences in timing of developmental change provide the core descriptive base for identifying agents of change and finding a principled basis for experimental interventions.

5. *Exploit the instabilities at transitions to identify potential control parameters.*—Recall that dynamic theory states that one or several of the components of a system can act as the control parameter to engender change. In principle, small changes in the control parameter can cause large developmental shifts. How can the control parameter be identified? One strategy, which we illustrate herein, is to map the developmental trajectories of several candidate control parameters with the time-dependent profile of the collective variable. This requires concurrent measures of performance in the same subject in contributing domains that the experimenter has good reason to believe contribute to the phase shift. For motor development, this might be measures of body build characteristics, sensory or perceptual abilities, specific practice, muscle strength, or motivational changes. Examples of subsystems that contribute to a shift in a cognitive task performance may be memory capacity, perceptual abilities, knowledge base, or motivational or social factors. It is important to consider potential control parameters that are nonobvious and seemingly more distantly related to the performance task. This requires imagination on the part of the experimenter. For example, it was a major insight to postulate that the onset of independent crawling—a change in the motor domain—might engender major cognitive, affective, and social phase shifts (see the review in Bertenthal, Campos, & Barrett, 1984). Ideally, measures of contributing subsystems should be independent of the collective variable task.

The goal, then, is to lay out the parallel developing subsystems in a manner analogous to that in Figure 6 above. Some subsystems may be eliminated from consideration. If they do not show incremental changes (either increases or decreases) at the phase transition of the collective variable, they are unlikely to have disrupted the ongoing stability. Other subsystems may emerge as more likely possibilities because they are changing rapidly at the same time as the phase shift. This stage of analysis is only correlational and suggestive. More conclusive evidence of a developmental control parameter requires experimental manipulation as discussed below.

In the present *Monograph*, the treadmill acts as a real-time control parameter that shifts the infant into a more mature coordinative form—alternate stepping. Our task was to identify what aspects of the infant-treadmill system were the control parameters that allowed this skill to evolve in developmental time as well. Our experimental design included concurrent measures of contributing subsystems to provide clues as to what made the infants change. Identifying putative control parameters allows for the next step, namely, their experimental manipulation.

6. *Manipulate the putative control parameters to generate phase transitions experimentally.*—In dynamic systems terms, points of transition allow the experimenter to test potential control parameters. That is, by identifying and manipulating those internal or external elements that shift the system into new behavioral configurations in real time, we will have clues as to what these elements are in developmental time. When behavioral attractors are stable (the ball is deep in the potential well), they are resistant to experimental perturbation. It would require an expert illusionist, for example, to elicit a nonconserving response from a normal 10-year-old child in a volume conservation task. When behavior is in transition, however (the ball is rolling around several shallow hills), small pushes from outside forces will easily move the system into one of several, more or less mature, configurations. Thus, even a conserving 5-year-old can be easily "fooled" into a nonconserving response by the appropriate task manipulation (Gelman & Baillargeon, 1983).

Developmental researchers have long exploited this characteristic of systems in transition—that they are easily perturbed—to design experiments that move these systems around in their developmental landscapes. The most direct test of a potential developmental control parameter is long-term intervention. For example, the Head Start program was based on two important developmental assumptions: that children learn by being exposed to a rich experiential environment and that they are especially vulnerable to this environment during the preschool years. Planners understood that cognitive abilities were undergoing rapid changes between ages 2 and 5 years, and they also reasoned that certain features of the environment fed these changes. In that Head Start did have long-term consequences for children of economically deprived backgrounds, the intervention did identify a point of transition (interventions in junior high would be less successful). The specific control parameter or parameters are less easily defined, however, since the intervention included cognitive, social, and nutritional enrichment. Indeed, all these likely acted together.

For obvious practical and ethical reasons, direct tests of developmental control parameters are rare. More commonly, researchers perform what are sometimes called "microgenesis" experiments (Vygotsky, 1962; see, e.g., Kuhn & Phelps, 1982; Siegler & Jenkins, 1989). These are attempts to mimic developmental changes by real-time experimental manipulations. These experiments typically supplement a task with explicit teaching or enrichment in order to push the child into more mature performance or impoverish the task so that the child reverts to an earlier response form. Microgenesis experiments work because the system is unstable and vulnerable and autonomously seeks new stable regimes. Nonetheless, the burden rests on the experimenter to show that the control parameter effective in the real time of the experimental manipulation is plausibly and believably the control parameter that shifts the system in developmental time. In this *Monograph,* we identify potential control parameters for treadmill stepping, but direct and indirect confirmation is left for ongoing work.

7. *Dynamic accounts at many levels of analysis.*—At this point, some may object that this strategy only pushes back a causal analysis to another level. Why stop the analysis with the identification of a control parameter? If changes in a contributing subsystem disrupt the stability of the collective variable, why not look for the multiple contributions to the changes in that subsystem as well?

We would respond that decisions about the level of appropriate causality must be made in any developmental study (or in any science). Some are content to ascribe a behavioral change to an anatomical change in the brain such as a new level of connectivity or a period of rapid synapse elimination. For example, Diamond (1990; see also Goldman-Rakic, 1987) has proposed that the successful solving of the "A-not-B" search task at about 9 months is due to the maturation of the prefrontal cortex. But is this the final cause? Others might ask what the neural or behavioral events are that engendered these changes in the brain. In a dynamic view, some control parameter must be changing in order for the stability of the existing organization to be disrupted and a new organization to appear. Thus, what is the explanation for some is the phenomenon to be explained for others.

In truth, of course, all these levels of explanation are important, interesting, and ultimately necessary, but dynamic systems differs from the more traditional views by not assigning causal privilege to any level. Thus, we do not seek a reductionist answer. Behavior is not reduced to brain structure, and brain structure is not reduced to DNA. Subsystems are equal contributors in principle, although the system stability may be more sensitive to some elements than others.

Because a dynamic approach is content free and based on very general principles of pattern formation, it can be, and has been, used at all levels of analysis from the level of neuronal activity to that of social systems. That is, the physiological or neural contributions to behavior can themselves be subjected to a dynamic analysis, as can the social environment. Glass and

Mackey (1988) give many examples of dynamic analyses of such diverse biological phenomena as autonomic nervous system function, hormone release, cardiac rhythms, respiration, motor tremor, pupil dilation, circadian rhythms, and cortical electrical activity. But dynamic principles have also been used to quantify the coordination of action between two people (Schmidt, Carello, & Turvey, 1990). Multiple, parallel, and interacting layers contribute to all behavioral performances and their developmental courses. Thus, a goal for a full understanding is not the reduction of one level to another but a description of the dynamics of all the participating levels and their integration.

In the next chapter, we introduce our translation of dynamic principles to a particular phenomenon of infancy—treadmill-elicited stepping. We suggest that traditional explanations of skill acquisition have not adequately accounted for the dynamic nature of the onset of walking, and we propose that this process can be illuminated by a dynamic analysis of one contributing component—the generation of alternating step patterns. This ability is manifest during infancy only with special environmental facilitation. We propose that, although this ability is necessary for walking, it is not sufficient. The developmental profile of this hidden motor behavior suggests that it is not the control parameter for the emergence of walking but a component that is masked by many other factors. We then offer speculations about the putative control parameter and suggest ways to confirm this possibility.

IV. DYNAMIC PROCESSES IN LEARNING TO WALK

In this chapter, we first propose that conventional neuromaturational or cognitive-maturational theories of learning to walk provide an inadequate account of a multidimensional, dynamic developmental process. We introduce the phenomenon of treadmill-elicited stepping as strong evidence in favor of a dynamic approach. We then outline our empirical study of treadmill stepping on the basis of dynamic principles.

WHY INFANTS WALK WHEN THEY DO

The first independent upright steps signal a discrete and dramatic "stage" in the developmental progression of forward locomotion: parents can usually remember well the moment their children took their first steps. Because walking onset is a clearly distinguishable developmental milestone, and one that is manifestly important for both infants and parents, it has been a classic topic of developmental study. For a long time, our views of locomotor development, and motor development in general, have been dominated by the work of the early pioneers of scientific developmental research, especially Arnold Gesell and Myrtle McGraw. These investigators described in exquisite detail the changes in posture and movement that preceded walking and other motor milestones. They are best remembered for their maturationist theories of development: that is, that new forms appeared by autonomous changes within the organism reflecting a phylogenetic blueprint. Developmental rate and direction could be somewhat modified by experience, but its fundamental plan could not.

Although these maturationist views soon went out of fashion for mainstream developmental studies, especially with the competing influences of learning theory and Piaget, they have continued to exert an important influence in the motor sciences (Thelen, 1987). Indeed, it is hard to overestimate the influence of McGraw and Gesell on conventional views of motor development as reflected in standard textbooks, on the construction of developmental tests for infants and children, and especially on theories of treatment for the physically delayed and disabled (e.g., Fiorentino, 1981). But equally important, we believe, is that the type of causal explanation used by early maturational theorists—ascribing behavioral change to structural changes in the brain—is also sometimes invoked in contemporary behavioral development. A dynamic systems approach fundamentally questions such explanatory principles for both locomotion and other developmental domains.

In her classic study *The Neuromuscular Maturation of the Human Infant* (1945), McGraw described the "phases of erect locomotion" as being determined by the increasing influence of the cortical areas of the brain on the subcortical and reflexive levels of neural functioning (Fig. 8). Thus, the step-like movements that infants perform during the newborn period are manifestations of "primitive subcortical nuclei" (McGraw, 1945, p. 10) that produce reflexive and phylogenetically old patterns such as stepping and "swimming" movements in a prone position. These patterns become progressively inhibited over the first few months as cortical inhibitory centers mature. The deliberate steps taken in phase D (Fig. 8) are evidence of the "onset of cortical participation," although cortical control of the process is not complete until walking acquires smooth and automatic adult-like qualities, including heel-strike and toe-off progression (new walkers land and take off on flat feet). Locomotion thus emerges as the nervous system matures in a hierarchical fashion, ontogenetically recapitulating the phylogenetic history of increasing corticalization of behavior.

It is useful to understand that, at the time of her pioneering work, McGraw was heavily influenced by contemporary advances in vertebrate neuroembryology, and especially the work of Coghill and others, that directly related changes in the central nervous system seen by histological techniques to behavioral changes manifested by developing motor patterns. The assumption was that material structure—in this case, the neuroanatomy—had a causal, privileged, and one-to-one relation to the behavior the animal displayed and that development in the former directed development in the latter.

McGraw herself acknowledged the complexities of the process of acquiring locomotion, recognized the dual interactions of structure and function, and admitted in the 1962 edition of her book that her efforts to seat development in the structure of the nervous system were not successful. Nonetheless, this line of "neurological causality" continues to have considerable appeal (see, e.g., Diamond, 1990; Fischer, 1987; Hofsten, 1989). It goes without question that the nervous system changes dramatically during postnatal development and that a complete understanding of behavioral development must include its neural basis. What is questioned is the assignment of final causality to central nervous system maturation.

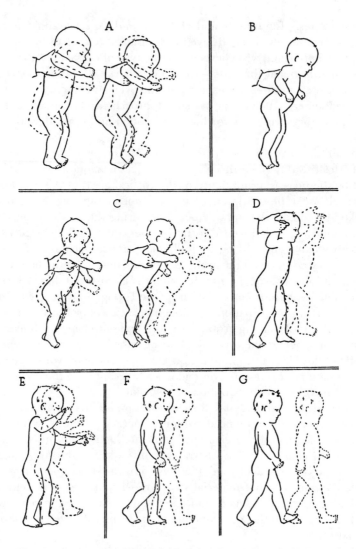

Fig. 8.—The seven phases of erect locomotion (from McGraw, 1945; used with permission).

The contemporary neuromaturationist view of locomotor development is reflected in the work of Forssberg (1985; Forssberg & Wallberg, 1980; Leonard, Hirschfeld, & Forssberg, 1988). This neuroscientist sees walking as the result of ontogenetic encephalization of primitive spinal pattern generators, much as did McGraw. Thus, the developmental process is "transformation" or "maturation of mechanisms" in the higher brain centers during infancy, which suppresses the primitive newborn walk and allows for mature features to emerge. In a recent paper, Leonard et al. (1989) admit that "walking is both a learned skill and one that reflects maturational changes of the immature nervous system" (p. 41). This statement cannot be disputed: it is, however, causally impoverished because it addresses neither what is learned nor how the nervous system matures.

More psychologically oriented neuromaturational accounts of locomotion have also been proposed. Several investigators have noted the correspondence between the onset of walking and other cognitive skills, such as first words and functional play. Zelazo (1976), for example, ascribed walking to the infant's newly found abilities to integrate movements and distal goals, a skill that emerges with a major cognitive reorganization at 1 year. Recent neuroanatomical studies give credence to the notion that major cortical reorganizations underlie these seemingly correlated new abilities. In the monkey, for example, Goldman-Rakic (1987) discovered that synaptogenesis—the formation, outgrowth, and indeed elimination of axons—occurs rather synchronously in all areas of the cerebral cortex, visual, motor, and associative (this was a surprising finding since these functions appear to mature serially, with sensory abilities appearing before higher association skills, etc.). According to Goldman-Rakic, these anatomical findings mean that "cardinal functions of each major area might emerge in at least elementary form in all regions of the cortex simultaneously. . . . Perhaps the coincidental timing of a child's first utterance and first step are expressions of concurrent synaptogenesis in the language and motor areas of the cortex" (p. 615).

Neuromaturational accounts of locomotor development are, by themselves, deficient on a number of grounds. First, there is the logical problem of assigning any particular level as the "cause" of the level above it. If rates of synaptogenesis "cause" cognitive development, one could always ask, What causes the synaptic density to change? There is equally compelling evidence that both synaptic density and patterns of connectivity are highly experience dependent (Edelman, 1987; Greenough, Black, & Wallace, 1987). The question may thus be rephrased, What do infants do in their first 8–12 months that produces specific timing and patterns of brain anatomy? In a dynamic approach, all these levels of explanation are important and worthy of study, but none are privileged.

Second, we know of no evidence to suggest that widely disparate skills such as the first utterance, first step, onset of functional play, or performance on a delayed-response task (Goldman-Rakic's metric of a "core cognitive ability") are coincidental in the same child. Indeed, these milestones are often separated by many months within an individual. Although neural remodeling, on either the cortical or the spinal level, may well occur, a developmental theory must account for this pervasive décalage in performance in a principled manner. Another objection relates to what gets identified as the "cardinal function" or "core capacity" (Goldman-Rakic, 1987, p. 616) of speaking or walking or indeed any particular task (see also Fischer & Bidell, in press; Thelen & Smith, in preparation). Why, for example, is the first step a more "cardinal ability" than the first crawl or creep? Why is a particular hiding task taken as the hallmark of "the ability of an organism to guide its behavior by stored information" (Goldman-Rakic, 1987, p. 602) when other tasks reveal that infants can guide their behavior by stored information many months earlier (Rovee-Collier & Fagen, 1981)? Similarly, how does one identify "maturation" in brain structure? Is a structure "mature" with a particular synaptic density or with a known level of a neurotransmitter? In a dynamic approach, in contrast, there is no "core" but rather distinguishable components that come together to accomplish tasks, using what abilities are available within a particular context. Thus, walking is not a pattern encoded in the spinal cord, and language is neither memory, symbolic representation, nor articulatory control. Both skills are a dynamic confluence of component abilities (Thelen, in press).

The main problem, however, with neurological maturation as a causal explanation is that it ignores the richness of the developmental process. While it is useful and important to know the patterns of synaptogenesis in the brain, this alone cannot tell us about the construction of real skills from precursor abilities that themselves have gradual or abrupt developmental timetables, in real children with dramatically different rates and levels of performance. It may be that a change in the synaptic density or some other structural component of the brain acts as the control parameter for a specific skill, such as the A-not-B error, but such a change is insufficient to understand the wide range of new skills acquired by infants between 8 months and 2 years. A single causal approach blinds the investigator to other sources of developmental change. The onset of crawling, for instance, triggers important changes in spatial cognition. In a sense, crawling is a nonspecific, noncognitive activity, yet it feeds back to engender cognitive transitions (Bertenthal, Campos, & Barrett, 1984). As we will detail below, the neuromaturational approach has been particularly inattentive to a number of nonneural and noncortical contributions to locomotor development.

Thus, as we have suggested previously (Thelen, 1985, 1986a; Thelen & Cooke, 1987; Thelen, Ulrich, & Jensen, 1989), the developmental milestone of learning to walk is not the result of a dedicated "locomotor switch" turned on somewhere in the central nervous system but the confluence of many relatively autonomous processes, each with its own developmental history. Many of these processes are functional and may be used in other contexts, long before the infant actually walks alone. For example, the motivational aspect of locomoting toward a desired goal is independent of the means of achieving that goal. Infants will scoot, roll, shuffle, creep, crawl, or use a wheeled device to reach a desired object often 6 months or more before they walk. It is not the translation of intention into motor action that limits locomotion but the developmental readiness of the sensory and neuromotor elements needed for particular postures and actions. In this *Monograph,* we document how one component ability, the alternating patterning of the legs, interacts with other developing components in the gradual construction of walking skill.

WHAT DOES IT TAKE TO WALK?

Upright bipedal locomotion, although seemingly our most automatic of skills, is really a highly complex and demanding motor task. First, a walker must generate a synchronized ensemble of muscle contractions to produce the locomotor movement. This usually involves muscles spanning many joints and body segments: the legs alternate in a pattern of swing and stance, the pelvis rotates and tilts, and the arms and shoulders swing forward and back in phase with the opposite leg (Inman, Ralston, & Todd, 1981). But locomotion, like all other motor actions, involves not just muscle contractions but also the interrelation between the motor patterns and the biomechanical and dynamic requirements of moving segments with mass and viscosity through a gravitational environment. For a biped to move forward, at least one foot must be airborne some of the time. For bipeds that walk rather than jump, the leg that is not airborne must bear all the body weight, and the weight bearing must alternate between the legs. The support leg needs to be strong and stable to support the body weight, and the animal must compensate for these alternating shifts of weight and remain balanced. This, in turn, requires the continual monitoring of the sensory messages of balance received from the visual system, the vestibular apparatus, and the soles of the feet to make the appropriate corrections within the changing biomechanical demands. Upright walkers must also be able to compensate very quickly for unexpected perturbations in paths, such as obstacles, uneven surfaces, and changes in direction (Gibson & Schmuckler, 1989).

How, then, do infants acquire this complex ability? We suggest that locomotion is constructed as a dynamic process, not imposed by maturational design. During development, there is striking autonomy not only of the motivational component, as mentioned above, but also of the processes contributing to the coordination of movement, posture, and balance needed to walk alone. For example, newborn infants are biomechanically unsuited for upright locomotion because they have large heads and trunks and proportionally small and weak limbs. Their high center of gravity means that controlling the pendular sway in stance is very difficult. During the first year or two, infants' proportions and body composition become more suited to bipedal locomotion. These are gradual, nonneural changes that nonetheless are as important to attaining independent gait as the patterning of the limbs (Thelen, 1984). Likewise, walking requires erect posture. Erect posture is also acquired gradually during the first year, in anticipation of locomotion. Posture is maintained through visual, vestibular, and somatosensory perception, and studies have shown that sensitivity to these modalities and the ability to integrate multimodal postural input develop asynchronously. For example, when 4-month-old infants seated in an infant seat were subjected to a rapid translation of that seat in a forward-to-back motion, they had poorly organized responses in the postural muscles of the trunk. However, when the same infants were deprived of vision with opaque goggles, they showed appropriate responses in the neck and trunk (for a review, see Woollacott, Shumway-Cook, & Williams, 1989). At 4 months, vestibular input elicited postural compensation, but vision plus vestibular sensations did not. When does a child "have" erect posture? Again, the question cannot be answered, but we can map a series of task-related responses and their relative stabilities.

The most dramatic demonstration, however, of this asynchrony of developing components was our discovery that 7-month-old infants who were supported on a motorized treadmill performed highly coordinated, alternating stepping movements with many adult-like qualities. These infants could not walk, stand, or even perform stationary stepping movements without the treadmill. It is this surprising "hidden" ability that we explore in detail here.

TREADMILL STEPPING

The nature of treadmill stepping in 7-month-old infants.—Thelen (1986b) compared the kinematic properties of leg kicks and steps of 7-month-old infants when they were supine, held erect with no treadmill, and supported on a treadmill moving at two speeds, fast (.19 m/sec) and slow (.1 m/sec). Leg kicks were included because spontaneous kicking is similar in coordination and temporal organization to erect movements seen in the newborn stepping response. Unlike the stepping response, which can no longer be elicited after about 2 months, infants continue supine kicking throughout the first year.

The treadmill dramatically facilitated stepping. Although infants performed a few step-like movements without the treadmill, the step rate increased manyfold when the treadmill was turned on. In addition, the percentage of steps that were strictly alternating increased from about 40% to about 85% with the treadmill.

The within-limb coordination patterns elicited by the treadmill were more mature than the patterns seen either in stepping without the treadmill or in supine kicking. In adults, the flexions and extensions of the hip, knee, and ankle joints are phased in a complex and precise manner. In newborns and young infants, kicks and steps show a much simpler organization: all the joints move nearly synchronously. Treadmill steps were much more like adult steps than were the no-treadmill steps and kicks in the same infant.

Although these 7-month-old infants appeared not to initiate treadmill stepping in any intentional or voluntary manner, the stepping response is more than a simple reflex. Infants responded to the faster treadmill speed by increasing their step rate, as though they were adjusting their overground traveling speed. They made these adjustments, as adults do, by decreasing the time the support leg contacted the ground while maintaining a nearly constant airborne or swing phase.

The infants were responsive not just to the overall speed of the treadmill but also to the interaction of the dynamic information delivered to each leg. Thelen, Ulrich, and Niles (1987) constructed a split-belt treadmill, where each of two parallel belts could be independently controlled. When confronted with conflicting speeds (one belt driven at twice the speed of the other), infants nearly perfectly maintained their alternating steps. They accomplished this by increasing the stance phase on the faster belt and concomitantly decreasing the stance phase on the slower belt. Stated another way, they began the swing of the leg forward at different points in the step cycle depending on the moving status of the other leg.

These results revealed a "hidden" ability of considerable complexity in 7-month-old infants. We do not know exactly how the treadmill elicits stepping movements. One likely explanation is that the mechanical pulling of the leg backward stretches the leg muscles and allows them to store energy, much like stretching a spring beyond its equilibrium point. When the leg is stretched to its anatomical limit, it uses this stored energy to spring forward. In mature locomotion, such a stretch in the trailing leg naturally results when the center of mass of the person moves over the forward leg as it contacts the ground. Indeed, biomechanists believe that this pendular swing that results from stored elastic forces is an important part of the step.

But a simple, mechanical explanation of the treadmill effect is not enough. We cannot account for the reliable alternation of the legs on the treadmill or for their sensitivity and adjustment to perturbations in bilateral symmetry. At some level, therefore, infants must be able both to detect that they are moving and to use that information in a functional manner. Since they received no motion clues from vision, they must rely on a combination of the somatosensory inputs from the soles of their feet and the proprioceptive messages about the status of their joints and muscles. Contact with the surface alone was not effective in eliciting steps; these receptors must detect dynamic changes suggesting motion.

One essential functional task in bipedal walking is to alternate the legs so that at least one leg is contacting the ground. (No matter what happens, one cannot be caught with both legs in the air!) It is significant that infants used the detection of motion to initiate alternation when the treadmill was turned on and to maintain alternation when the legs were perturbed. Infants, therefore, transduced sensory information into motor activity that will have important functional consequences 4–6 months later, and they did this the very first time they encountered the treadmill and with no apparent voluntary effort.

A sophisticated perception-action system for locomotion is in place, therefore, long in advance of the normally performed behavior. The perception system is tuned to the dynamic context, and the motor system coordinates the contraction of muscles in both legs in a very precise way. The fundamental unit in this system is a two-leg synergy, which is context sensitive as a functional ensemble.

The observation that at 7 months infants can step on the treadmill but not without it, yet use their legs for locomotion in creeping and crawling and for other voluntary tasks, is compelling evidence that upright locomotion is the result not of a maturational "switch" but of the construction of a skill from continuous, available precursors. We use a dynamic analysis to understand one of those precursors.

Ontogenetic orgins of treadmill stepping.—The discovery of this hidden skill in 7-month-old infants opened intriguing questions about its developmental origins. Was this ability something infants are born with? If not, then when during the first 7 months could it be elicited? Was performance reliable within and between ages? Did this ability appear suddenly, or did it manifest gradual onset and improvement? Were there large interindividual differences in the course of treadmill-elicited stepping? Was this ability correlated with other known developmental milestones? How did treadmill stepping change as infants approached the age of independent locomotion?

Studying treadmill stepping from a dynamic perspective.—To answer these questions, we adopted an approach explicitly derived from dynamic systems in action and development. The first requirement in a dynamic strategy is to identify one or a few variables that capture the cooperativity of the many processes and subsystems that produce patterned behavior, the collective variable. As our primary concern is change over time, the next requirement is to describe, by longitudinal design, the developmental trajectory of the collective variable. This is done to identify places in the trajectory where new forms appear and where the system can be manipulated to uncover control parameters. Pooled group data, by averaging changes, are likely to mask these sensitive transitions. Thus, we elected to follow a relatively small number of infants longitudinally with the hope that normal individual profiles would offer clues to the underlying processes. We expected individual differences but maintain that these cannot be ignored as mere noise or deviations. Variability is as much a part of the developmental story as is uniformity. A satisfactory developmental explanation must deal with both the general processes at work in all human development and the variability that is an equal partner to these broad ontogenetic principles. Because dynamics deals with points of stability and instability, we also sought to capture performance variability within individuals. Thus, we repeated our exact test protocol within 1–3 days of the first session at each age. This also allowed an assessment of best performance, as infants are notoriously capricious in laboratory settings.

Our experimental design exploits dynamics on two time scales. In addition to the major goal of describing shifts in treadmill stepping between ages, we also incorporated a scalar within each age. These manipulations were inspired by an important and growing body of literature on adult voluntary cyclical movements, including locomotion, but especially dealing with bimanual arm and hand movements (Kelso, 1984; Kelso & Scholz, 1985; Kugler & Turvey, 1987; Schöner & Kelso, 1988). Very briefly, these studies show that the coordinative patterns of bilateral entrainment of the phasing of hands or legs is a function of the cycle frequency (a measure of the "energy" delivered to the system). Most important, the system appears suddenly to "switch" coordinative modes at certain individually stable points on this frequency metric: subjects asked to speed up 180-degree out-of-phase hand and finger movements shift into an in-phase mode as they move faster. In dynamic terms, the system shifts from one cyclic attractor to another as the first movement pattern becomes unstable at increased speeds (Kelso, Scholz, & Schöner, 1986). We were likewise curious to see if these involuntary infant movements would show similar regions of instability and phase shift into new coordinative modes by testing stepping at gradually increasing treadmill speeds.

We tested system stability finally by introducing two kinds of perturbations. First, we increased the treadmill speed within an ongoing test trial. The change of treadmill speed acted to disturb the ongoing activity. If the attractor state were stable, we would expect the infant to correct immediately by maintaining alternating steps, while a less stable state would be more readily disrupted. Second, we repeated the "split-belt" treadmill challenge at each test session. This is a measure of how well the infant keeps alternating steps despite the extreme perturbation of having one leg moving at twice the speed of the other. Remember that, by 7 months, infants compensate for the split belt with great precision and reliability.

These manipulations were designed to trace the entire course of treadmill stepping in the first year and to identify times when the system was undergoing rapid change. The ultimate goal of a dynamic approach to development is to discover the ontogenetic control parameters that drive the system into these new modes. This requires experimental manipulations at times of transitions. Although these manipulations were not performed in the present study, we did seek to identify potential control parameters to be tested. To do this, we videotaped all sessions to look for postural and behavioral indications of instability post hoc—that is, after we had identified the transition periods. In addition, we collected a variety of anthropometric measures since changes in body build and composition have been previously implicated in leg movement coordination. Finally, we assessed motor development independent of treadmill stepping by administering the motor portion of the Bayley Scales of Infant Development monthly.

V. METHODS

SUBJECTS

We identified 13 infants (seven girls and six boys) from published birth announcements and recruited them by an introductory letter and followup telephone call. Four of the recruited infants did not complete the study. We stopped testing two of these because they cried during testing; the parents of the other two chose not to continue after the first day of testing. Therefore, the data we present are based on the nine infants who continued as subjects through at least month 7. All babies were white, full term, and born without any known disabilities. Average birth weight was 3.794 kg (8 pounds, 6 oz). We paid all subjects for participating in each test session.

PROCEDURE

Each baby came into the Infant Motor Development Laboratory at Indiana University twice each month for testing. Testing began at month 1 and continued through month 7 for nine babies (with the exception of ES, who was not tested at month 5); additionally, two infants continued through month 8, three through month 9, and two through month 10. We stopped the sessions when the infant complained or refused the task consistently, which was usually accompanied by increased attention to the feet and to the treadmill. The two monthly test sessions were held within 2–3 days of each other and within 1 week of the monthly anniversary of the baby's birth. We asked parents to schedule appointments at times when their babies were usually awake and alert, that is, after naps and feeding times.

In order to help the infant adjust to the laboratory environment, each session began with 5–10 min of informal interaction between the experimenters, the parents, and the infant. On the first test day of each month, we followed this by an assessment of the infant's level of achievement of developmental motor milestones by administering the motor items from the Bayley Scales of Infant Development (Bayley, 1969).

Next, we prepared infants for treadmill testing by removing their clothes and attaching infrared emitting diodes (IREDs) to the lateral surface of the malleolus and fifth metatarsophalangeal joints of the right and left foot. We put lightweight socks on their feet that had holes cut out for the IREDs, so as to keep the IRED wires from becoming tangled as the babies moved and to make their foot contact with the treadmill belts as comfortable as possible.

The treadmill was constructed with two adjacent, parallel 87 × 16-cm belts, each driven by a variable speed motor. We placed the treadmill on a table, resulting in a height from floor to belt surface of .87 m. Eleven 20-sec treadmill trials were administered in each test session. During each trial, a research assistant supported the infants under their arms and in an upright position so that the infants' feet rested on the belts of the treadmill and they supported an apparently comfortable amount of their own weight. Because we believed that it was important for infants to support as much of their own weight as possible, we had to provide postural support that was sensitive to within- and between-baby variability. Therefore, we chose to provide human support rather than a less sensitive mechanical device. Trials 1 and 9 were baseline trials in which infants were supported on the treadmill but the belts remained stationary. During trials 2–8, both belts moved at the same speed, starting at a slow speed (.11 m/sec) for trial 2 and increasing speed in equal increments with each successive trial to reach .29 m/sec by trial 8. Our small sample size and infants' tolerance limits did not allow more than one trial order. Therefore, we chose a trial sequence of gradually increasing speeds to be consistent with previous investigations of shifts to new coordination modes during cyclic motor behavior. The belt speed was increased 5 sec after the trial began. Trials 10 and 11 were mixed-speed trials. In trial 10, the right belt moved at .11 m/sec, and the left belt moved at .23 m/sec (i.e., right belt slow, left belt fast); in trial 11, the belt speeds were reversed (i.e., right belt fast, left belt slow). We provided a break between trials 5 and 6 and between trials 9 and 10 that averaged 1.57 min. At the end of each trial, we recorded the infant's state of arousal according to the following code: 1 = drowsy, 2 = alert and quiet, 3 = alert and active, 4 = fussy, and 5 = crying.

We used the Waterloo Spatial Motion Analysis and Retrieval Technique (Watsmart), an opticoelectronic motion analysis system, to monitor right and left foot movements during treadmill testing. This system uses infrared sensitive cameras to detect individually sequenced pulses of light emitted by the diodes as point sources in a high-resolution grid (resolution = 2 mm at 4–6-m camera-to-diode distance, according to the manufacturer). The IREDs were pulsed at 100 hertz and tracked by two cameras positioned on

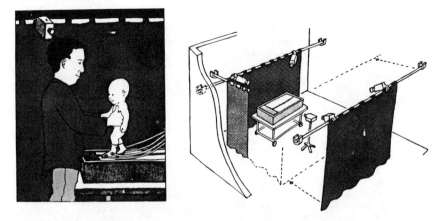

FIG. 9.—Artist's drawing of subject on treadmill and the laboratory setup

each side of the treadmill. Vertical height of the cameras was 2.15 m, horizontal distance from the treadmill was 1.38 m, and camera pairs were positioned 70 degrees apart relative to each other. The x and y coordinates from the separate cameras were subsequently converted to three-dimensional coordinates on the basis of calibration values established prior to each test session and were filtered using a low-pass (7 hertz) numerical filter via specialized hardware and software on an IBM PC-AT. In addition, all test trials were videotaped using a Panasonic color videocamera, recording at 30 frames per second, and a Sony Super Beta videocassette recorder. The videocamera was positioned on the infant's left side (see Fig. 9).

After treadmill trials, and on the first test day of each month only, we took a series of anthropometric measurements: weight, length from crown to heel and crown to rump, circumference of thigh and calf, and skinfold thickness at subscapula, triceps, and umbilicus. To determine reliability between the two experimenters responsible for all anthropometric measurements, one experimenter took all measurements of each of six infants on one day, and the following day the second experimenter repeated all measurements of the same infants. The range of Pearson product-moment correlations for these eight measurements was from .62 (thigh circumference) to .99 (weight), with a mean of .88.

DATA REDUCTION

The initiation and duration of step cycles and of the swing and stance phases were determined by directional changes, anterior and posterior, of the IRED on the base of the toe for each leg. If this IRED was obscured

49

TABLE 1

LEG AND FOOT POSTURE CATEGORIES

Baseline trials.	1. Leg posture (flexed, extended)
	2. Foot orientation (toes pointed away from the mid-sagittal plane, toes toward mid-sagittal, foot in a sagittal plane)
Moving-belt trials[a]	3. Leg posture (flexed, extended)
Moving-belt trials[b]	4. Orientation of swing foot (toes pointed away from the mid-sagittal plane, toes toward mid-sagittal, foot in a sagittal plane)
	5. Degree of flexion in swing leg (high, low)
	6. Foot contact at touchdown (toes, heel, flat, lateral side)
	7. Part of foot contacting surface during stance (toes, heel, flat, lateral side)

[a] Time samples in which the infant was not stepping.
[b] Time samples in which the infant was stepping.

for portions of a trial, the ankle IRED was used instead throughout that trial. Step cycles began with a swing phase, which we defined as occurring when the foot reversed its direction from being drawn backward by the treadmill belt to moving forward. Initiation of the stance phase occurred when the foot reversed direction from forward to backward, that is, approximately when the foot contacted the treadmill and began to be moved backward by it. Phase lags between step cycles of opposite legs were calculated as the elapsed time between the initiation of swing in the leading foot and the initiation of swing in the trailing foot divided by the duration of the cycle of the leading foot.

Because not all movements of the foot constituted alternating steps, we first identified steps from plots of the trajectory of foot movements in the anterior-posterior direction and then looked at our video record of the infant's movement for confirmation. A foot movement was coded as an *alternating* step if the step was initiated within 20%–80% of a step cycle on the opposite leg. Steps that were initiated before 20% or after 80% of the opposite leg cycle were coded as *parallel* steps. A *double* step occurred if within a sequence of alternating steps a second or extra step was taken by one leg without a concomitant second step by the other leg. A *single* step occurred when a step was produced by one leg without a step cycle on the other leg that overlapped it in time. Stepping was frequently interrupted during the 20-sec trials by periods of spontaneous flexions, extensions, rotations, air kicks, and inactivity. At this level of data reduction, these periods of activity and inactivity were coded only as nonstepping behaviors.

We coded an additional level of description of the infants' treadmill behavior as recorded on videotape by coding leg posture and orientation at 1-sec intervals throughout the entire trial for trials 1–9. We coded seven behavior categories and their variations, as outlined in Table 1.

Two coders analyzed the videotapes; percentage interrater reliability for each category was determined by dividing the total number of behavior variations seen by both observers by the total number seen by both plus the number seen only by observer A plus the number seen only by observer B. Mean interrater reliability based on the independent coding of 180 time samples, constituting nine trials for one infant, was 74%. The range across the seven categories was from 67% to 84%.

Except where indicated, all data presented are based on infants' performance on their better of two test days within individual months. We defined "better" day as the one that had more alternating steps.

We have organized our report of the results to reflect a dynamic strategy for understanding the development of treadmill stepping. First, we identify a collective variable. A large portion of the results is then devoted to the second task, namely, describing the dynamics of that collective variable over two nested time scales: ontogenetic time and within the speed scalar of the experimental manipulation. Third, we report various measures of stability and instability, including within- and between-subject variability and responses to perturbation. Finally, we explore various candidates for developmental control parameters among our measured dependent variables.

DEFINING A COLLECTIVE VARIABLE

The hallmark of task-specific treadmill stepping is the emergent ability to alternate the legs in walker-like fashion when the mechanical action of the treadmill pulls both legs backward. Alternation requires an informational link between the legs such that the dynamic condition of one leg is used to regulate the initiation and trajectory of the second leg. Single, parallel, and double steps may be purely mechanical, that is, determined strictly by the potential energy stored by the stretched leg and the spring-like action of the limb when released. Thus, selecting alternating steps as a dependent variable captures the cooperative ability of the system to respond to the imposed task. This measure condenses the contributions not only of the biomechanical elements of the infants' skeletomuscular systems at any particular maturational level and the neurologically determined tension in the muscles but also of the neural transduction between the biomechanical demands and the infants' motivational states.

We report several metrics of alternating steps. For the entire sample of nine infants, we recorded overall number of alternating steps, which estimates the infants' performance on the treadmill as a function of age and

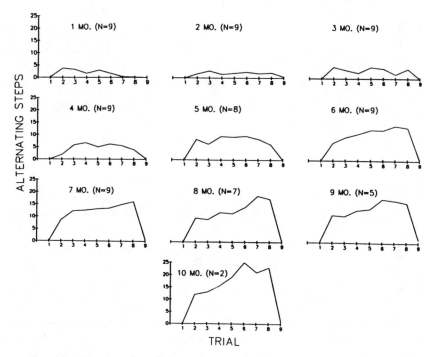

Fɪɢ. 10.—Mean number of alternating steps by trial and age, pooled across all subjects. Trials 1 and 9 are baseline trials. Unless noted otherwise, all data are for the better performance day.

task. For a subsample of four infants, we extracted more detailed kinematic variables, namely, the cycle durations of alternating steps, a measure of the timing of the movements, and the phase lags between successive steps, which quantifies the interlimb coordination.

THE DYNAMICS OF TREADMILL STEPPING

Age and Speed Effects on the Performance of Treadmill Steps

Figure 10 summarizes the pooled data for the average number of alternating steps performed by the nine infants at various ages and treadmill speeds. At all ages, the infants took very few steps in trials 1 and 9, when the treadmill was not moving, but stepped when the treadmill was turned on. In general, there was an increase in the number of steps with age and with increasing treadmill speed. The speed effect was especially apparent

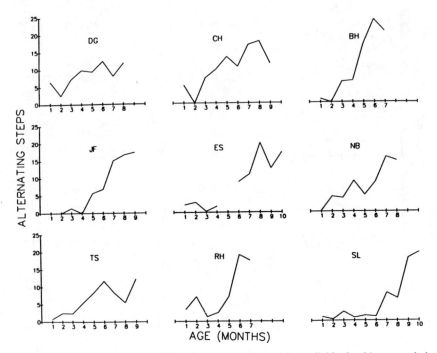

Fɪɢ. 11.—Mean number of alternating steps by age for individual subjects, pooled over speed trials.

in trials 2–6 and in the second half of the first year. The general linear model for regression analysis was used to run a repeated-measures ANOVA. The statistical tests for both linear and quadratic age-related trends collapsed across trials 2–8 were significant, $F(1,47) = 83.44$, $p < .001$, and $F(1,47) = 4.69$, $p < .035$, respectively.

Since the pooled sample data conceal important individual similarities and differences, the effect of age on treadmill stepping is illustrated in Figure 11 for the individual infants, reported for their better performance day and pooled over trials. A number of infants took steps on the treadmill in their first month. Three (DG, CH, BH) showed rapid and nearly linear increases in stepping following month 2 (the *early steppers*), and all showed some decline in performance during the last few months. Infants JF, ES, RH, and SL, in contrast, performed relatively few steps in the first 3 or 4 months (the *later steppers*) and increased their output quite steeply thereafter. Several of these infants also showed some decline in their stepping in months 7–10.

Despite these differences, the nine infants displayed a similar pattern in their developmental profiles of alternating steps: an initial period with

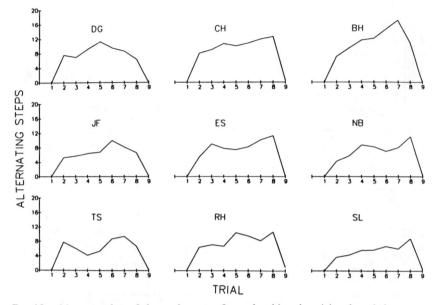

ALTERNATING STEPS

TRIAL

FIG. 12.—Mean number of alternating steps for each subject, by trial and pooled over age

only a few steps, followed by a short period of decline in steps and then a more steady increase lasting several months. For most, the increase was followed by a period near the end of testing during which stepping declined.

The individual infants' responses to speed changes in the treadmill are presented in Figure 12, with data averaged over ages. Again, over all ages there was a general tendency to increase the number of steps when the treadmill moved at a faster rate, although for many infants the highest speeds led to a decline in steps. Again, the general linear model for regression analysis was used in a repeated-measures ANOVA. The statistical tests for speed effect across trials 2–8 (collapsed across age) were significant for both linear and quadratic trends, $F(1,48) = 26.19$, $p < .001$, and $F(1,48) = 4.16$, $p < .046$, respectively.

This interaction between age and treadmill speed on the number of alternating steps is best seen in the individual developmental profiles (Fig. 13). Infants consistently took no or few steps in the baseline conditions either before or following the treadmill trials. Infant DG, who stepped the most during the early months, had his peak step rate at moderate speeds (trial 5) in months 1–5 but shifted to prefer moderately fast speeds (trial 7) in months 6–8; at all ages, the fastest speeds were less effective for him (Fig. 13). No other infants made such consistent speed-related adjustments before months 4 or 5, but by 6 months all were taking more steps as the

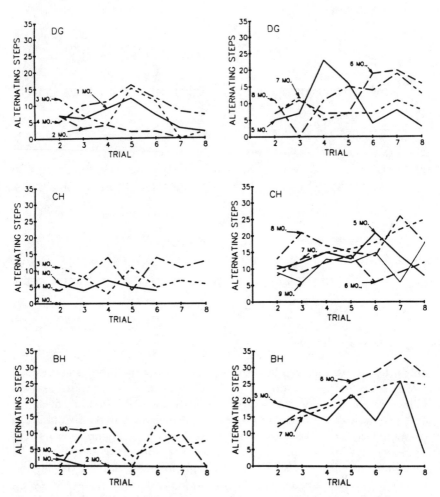

FIG. 13.—Number of alternating steps for each subject by age and trial

belt speed increased. Commonly, the moderately fast speeds elicited more steps than the fastest belt speed of trial 8. Seven of the nine infants showed a decrease in the overall number of steps in the last month of testing. This is not surprising since cessation of testing was (in part) contingent on the infant's competing behaviors or lack of interest in the task.

In sum, the treadmill elicited at least some steps in nearly all the infants during the first few months of life, and all were stepping well by months 4 or 5. Four appeared to begin increases in stepping by month 3; the remainder showed their dramatic increase at month 5. All the infants showed consistent adjustment to the speed changes in the treadmill by the second half of the year; infant DG responded to treadmill speed even at month 1. Nearly all

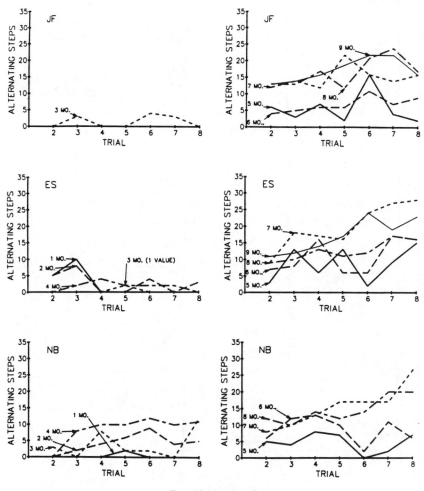

FIG. 13 (*Continued*)

the infants performed their highest step rate at less than the fastest belt speed; however, the design of the experiment makes it impossible to know whether this reflects simply fatigue or an optimal match between belt speed and some naturally preferred stepping frequency.

Developmental Changes in Step-Type Repertoire

Recall that a step could occur within a number of different movement sequences in addition to being followed by a step on the opposite leg. The step could be simultaneous with the step on the opposite leg (parallel), or followed by a step by the same leg (double), or not followed by another step

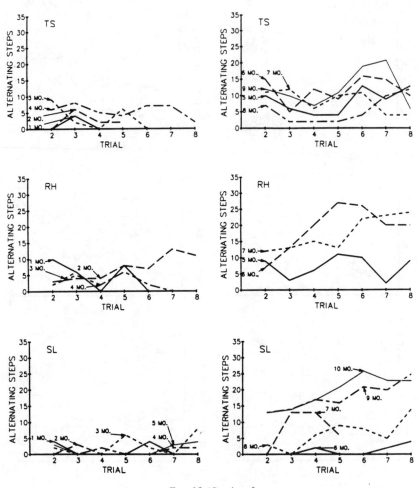

FIG. 13 (*Continued*)

at all (single). Note also that the infant had the option of not moving at all or of moving in other non-step-like patterns; thus, the total number of alternating steps is independent of their proportion in the trial (this is not strictly true for instances where the infant stepped for the entire trial).

Figure 14 reports the monthly proportion of the total number of steps that were classified as alternating, single, parallel, or double. What is note-worthy about these graphs is that the stepping repertoire becomes increas-ingly dominated by alternating steps. With the exception of infant DG, our best and earliest stepper, who had a high proportion of alternations right from the start, each infant had an initial period characterized by variety of

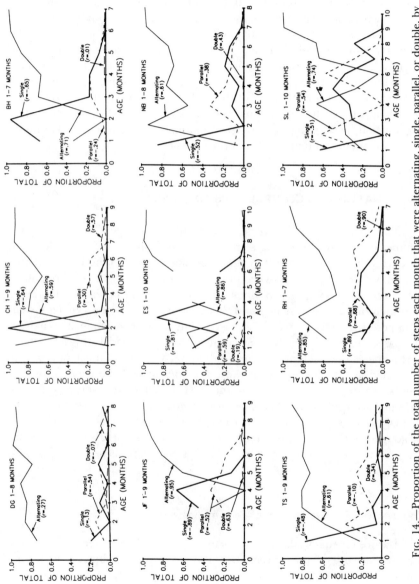

FIG. 14.—Proportion of the total number of steps each month that were alternating, single, parallel, or double, by individual subject and pooled over trials. The r values equal the correlation between absolute number of steps taken and proportion.

step types and then "settled in" to a high proportion of alternating steps. If we compare Figures 14 and 11, we see that for each infant the increase in the absolute number of steps is accompanied by a decrease in their variety. Correlations between absolute number and proportion of alternating steps over all ages and trials were as follows: DG = .27; CH = .59; BH = .71; SL = .74; JF = .95; TS = .61; NB = .61; ES = .86; and RH = .85. Consider infant SL, for example. She did not show a dramatic increase in the number of alternating steps until month 6. Before that time, she performed a variety of step types even though, at least in principle, she might have taken only a few steps, all of them alternating. When her step performance improved, her repertoire became strikingly reduced. These data suggest that, as stepping on the treadmill increased, alternation became a preferred mode compared to the other possible step types.

COORDINATION AND CONTROL IN TREADMILL STEPPING

The analysis reported above showed that infants performed a greater number of strictly alternating steps on the treadmill as they got older and, concurrently, more steps when the treadmill speed was increased. In this section, we ask about improvement in the movement quality of those alternate steps by looking at two kinematic measures that index the degree of coordination between the legs and the responsiveness of the two-leg system to imposed changes of treadmill speed. These detailed kinematic analyses are based on every strictly alternating step in every trial on the "better" performance day each month for a subsample of four infants ("better" day is defined as the one in which the infant performed the greatest overall number of steps). Three of the infants, CH, DG, and BH, were in the early stepping group (see Fig. 11); they are interesting because they allowed us to analyze the structure of the earliest steps and, presumably, to observe a prolonged period of improvement. The fourth infant, JF, was drawn from the later stepping group (see Fig. 11) so as to contrast his performance with that of the earlier steppers.

Coordination: Relative Phasing of Bilateral Steps

Mature bipedal walking is characterized by bilateral symmetry in movement: the two legs move precisely 180 degrees out of phase. This means that walkers begin their second step halfway through the cycle of their first step. (Other gait patterns not involving strict alternation such as hopping, galloping, and jumping are possible but are universally not preferred by humans for ordinary forward locomotion.) The precision of bilateral alternation is a measure of the coordination between the legs. Information about the dynamic status of one leg is used to control the movement of the opposite leg so that the timing of their phases is precisely maintained.

When the treadmill mechanically pulls the legs backward, their alternating response means that such an information system must be in place, whereby one leg's spring forward is inhibited until the correct phase of the movement of the other one. When does this ability develop?

Recall that the relative phase lag was determined as the time between the initiation of the swing phase of the treadmill step in the leading foot and its initiation in the trailing foot. Because step cycle durations are quite variable (see below), each phase lag was normalized to a proportion of the step cycle by dividing it by the cycle duration:

$$\frac{(\text{Time of step onset trailing foot}) - (\text{Time of step onset leading foot})}{(\text{Cycle duration of leading foot})}.$$

This means that perfectly alternating steps would have a phase lag of .5. Figure 15 summarizes the age changes in relative phase lags for the four infants and in the normalized standard deviations of individual phase lags, pooled over all the trials. Each of the four infants more closely approached the ideal .5 lag after month 4; by age 6 months, infants had stabilized or were steadily decreasing the variability in their phase lags. Mean lags below .5 throughout the year suggested that the trailing leg generally initiated stepping earlier in the cycle than in mature unsupported walking.

These pooled data mask more complex trends. Figure 16 reports the mean relative phase lags of all right- and left-leg treadmill steps of the four profiled infants (on their better performance day) as a function of age and treadmill condition. In the early-stepping infants—BH, DG, and CH— there is a clear trend toward a .5 phase lag, with less variability during the second half of the first year across all treadmill speeds. It is evident that, although steps alternate in the first months (by our definition that initiations occur within 20%–80% of the cycle of the opposite leg), they are nonetheless poorly coordinated.

These plots also suggest that, just as certain speeds appeared to elicit more steps, there is also an optimal treadmill speed for eliciting good coordination. The data portrayed in Table 2 report the standard deviations of the phase lags, collapsed across all months, as a function of treadmill speed. It can be seen that coordination stability is maximum for CH at the fastest three trials, especially trial 6, which also appears to be most facilitating for DG. If we eliminate the first four months, when overall variability is high, the speed effect becomes especially dramatic. Another noticeable trend is the consistent decrease in variability following the first moving-belt trial. Either the speed itself was more difficult for infants to respond to, or an

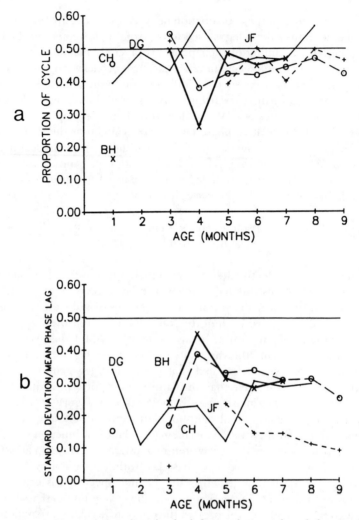

FIG. 15.—*a*, Mean relative phase lag by infant and age and pooled over trials. *b*, (Standard deviation)/(Mean phase lag).

adjustment period was required before the system settled into a more stable response cycle.

Control of Stepping: Cycle Durations and Speed Responses

In mature overground locomotion, the duration of the step cycle (the time between the initiation of a step and the initiation of the succeeding step on the same limb) depends on the speed of forward motion. Increasing

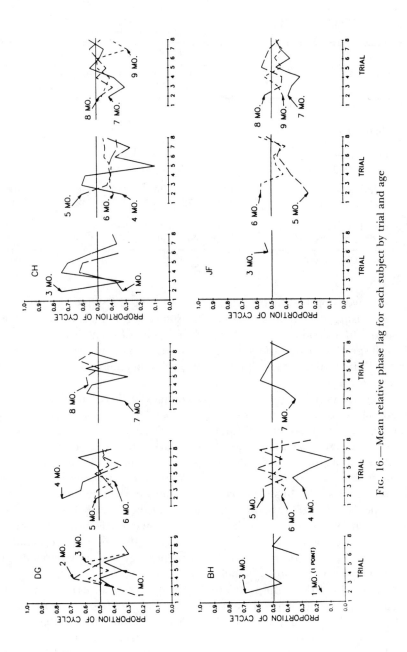

FIG. 16.—Mean relative phase lag for each subject by trial and age

TABLE 2

STANDARD DEVIATIONS OF THE PHASE LAGS BY SUBJECT AND TRIAL
ACROSS ALL AGES AND EXCLUDING MONTHS 1–4

	TRIAL (Belt Speed)[a]						
INFANT	2 (.11)	3 (.14)	4 (.17)	5 (.20)	6 (.23)	7 (.26)	8 (.29)
DG:							
Across all ages...............	.149	.088	.093	.108	.081	.127	.101
Excluding months 1–4........	.100	.078	.075	.109	.089	.097	.050
CH:							
Across all ages...............	.137	.087	.126	.168	.063	.091	.057
Excluding months 1–4........	.077	.049	.042	.054	.038	.089	.064
BH:							
Across all ages...............	.181	.087	.087	.140	.150	.091	.104
Excluding months 1–4........	.090	.072	.074	.067	.032	.088	.117
JF:							
Across all ages...............	.118	.089	.082	.055	.055	.052	.053
Excluding months 1–4........	.118	.093	.082	.055	.055	.037	.053

[a] Belt speed is given in m/sec.

speed is accomplished by progressively shortening the step cycle, primarily by decreasing the amount of time the stance leg is in contact with the ground. On a treadmill, the belt speed determines locomotion "speed." As the belt speed increases, the stance leg is stretched back more quickly. Adults on a treadmill match their (presumably) voluntary movements to the belt speed.

How do infants respond to speed changes in the treadmill? Is infant treadmill stepping a product of a belt-independent pattern generation that operates at a preferred speed? Or are treadmill step rates determined solely by the external driver, the belt? We saw earlier that increased belt speed produced more steps, especially in the second half of the first year. This suggests, but does not necessitate, that infants are stepping with shorter duration cycles and thus producing more steps in the same trial time. Another possibility is that the increase in steps at the faster belt speeds was a result of simply more steps, without any decrease in cycle duration. Remember that, since the infants do not have to support their weight and maintain their balance, there are no serious consequences for moving independently from the belt or not moving at all.

Figure 17 reports the individual cycle durations as a function of the treadmill speed and of age. Infants are indeed responding to the increasing belt velocity by taking quicker steps. Each of the infants shows speed sensitivity in step duration, even in the earliest months in which they stepped, but a clear trend toward a linear relation between speed and step time emerges only with increasing age. As in the case of simple number of steps,

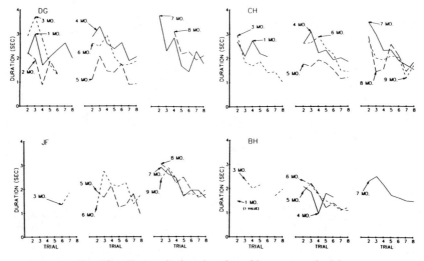

Fig. 17.—Step cycle durations by subject, age, and trial

the faster speeds appear to elicit more consistent step-cycle durations at all ages. Over all ages, the belt speeds used in this study elicited individual cycles averaging about 2.75 sec at the slowest to about 1.5 sec at the fastest speeds for infants DG, CH, and JF; infant BH stepped with quicker steps at all ages. Nonetheless, there was considerable variability both between infants and in the same infant over different months.

This variability in cycle duration showed one surprising age effect, however. In Figure 18, we present the average durations for all trials for each of the infants: for each leg, five of the eight minimum mean values occur at month 5. If the first month of stepping is not considered, all eight possible trajectories show this 5-month minimum. Thus, in this sample, infants are taking faster steps across speeds. This is independent of the overall number of steps, which continues to increase throughout the second half of the first year. We do not have an explanation for this speed increase.

There is clear evidence that even in the early months infants adjust their cycle duration to the speed of the treadmill and that this ability improves with age. Although cycle durations range between 1.5 and 3 sec, there is considerable variability in the actual cycle time. Thus, we can conclude that there is improvement in either the sensitivity to the moving treadmill or the ability to adjust to the speed by initiating the step, or both. There is suggestive evidence that there may be individual effects on "preferred" step cycle durations in the consistently more rapid steps of BH and the

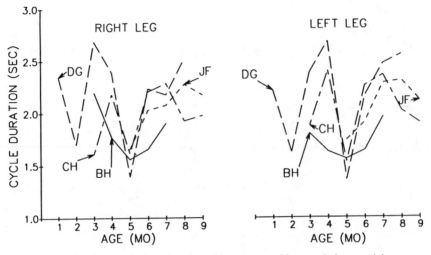

Fig. 18.—Step cycle durations by subject, age, and leg, pooled over trials

dramatic age effect in all the infants. In the next section, we characterize these preferred states in dynamic terminology.

TREADMILL STEPPING AS A STABILIZING ATTRACTOR

Our observations showed that, with age, infants stepped more on the treadmill, their repertoire became increasingly dominated by alternating steps, their steps were more sensitive to the belt speed, and the coordination between the two legs became more symmetrical. In short, the infants increasingly "settled in" to a preferred motor regime while on the treadmill; alternating stepping is a behavioral attractor state for infants when placed on a treadmill.

Dynamic systems theory predicts that, when a system is in such an attractor state, it will be both less variable and less vulnerable to disruption by perturbations. In this section, we show that this stabilizing attractor is indeed reflected in less variable performance.

Attractor Stability

Measures of Variability

Stability in within-month performance.—Infants in this study were tested on the treadmill on two visits, scheduled within a day or two. A stable system

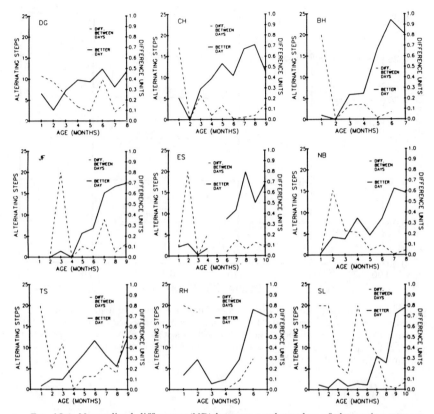

Fig. 19.—Normalized difference (ND) between total number of alternating steps on 2 test days within the same month and mean number of alternating steps on the better day. Both plotted by age and subject. ND = (Day 1 − Day 2)/(Day 1 + Day 2).

would be reflected in similar levels of performance on the 2 days of identical testing. Figure 19 illustrates the normalized difference between the total number of alternating treadmill steps on the first and second day of testing and the number of alternating steps on the better day, plotted as a function of age for all nine infants. In each case, there is a clear decrease in between-day variability as overall stepping skill increases. This is especially noticeable in "late" steppers TS, RH, and SL, whose large and persistent between-day variability sharply declined just as their performance began to improve dramatically. In a number of infants—JF, NB, DG, ES, SL, and to some extent

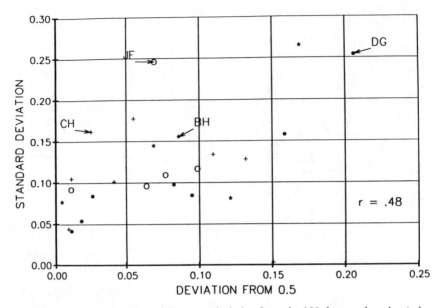

Fɪɢ. 20.—Absolute value of the mean deviation from the 180-degree phase lag (relative phase lag of .5) vs. the standard deviation of the phase lags for each subject's optimal trial age. Optimal trial defined as the trial with the most steps. Note that data points were graphed for each month in which a sufficient number of alternating steps were taken to determine a standard deviation.

TS (RH is missing the second day at the transition point)—the variability is the highest just before the sharp transition to the improving stepping total.

Stability in bilateral coordination.—We stated that mature walkers consistently maintain a .5 phase lag between the step cycles of each leg and that the infant systems also settled into this configuration with increasing age and, to some degree, at optimal treadmill speeds. Assuming that the strictly alternating state is the preferred attractor regime, we asked whether deviations from this coordination pattern were also reflected in increased variability. In other words, when infants were more consistently approaching the strictly alternating pattern, were they also stepping in a more consistent manner? To answer this, we chose the apparently optimal treadmill speeds (defined as eliciting the greatest number of steps) for the four profiled infants and plotted, for each month, the absolute value of the mean deviation from .5 against the standard deviation of the phase lags for that trial (Fig. 20). Whether the deviation was positive or negative, it is clear that, the closer to the ideal .5 coordination, the less variable the coordinative pattern. The correlation between these two variables was moderate and significant, $r = .48$.

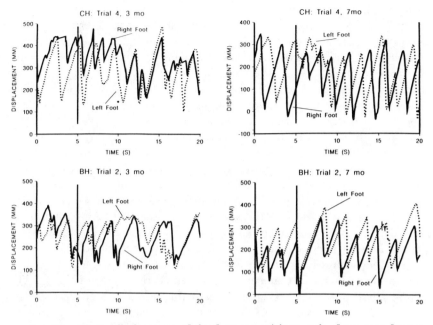

Fig. 21.—Linear displacement of the foot; two trials at each of two ages for two subjects. Increasing displacement values indicate that the foot is moving backward on the treadmill (stance); decreasing values represent the forward swing of the foot. The speed increase was introduced at 5 sec.

Perturbations

A second metric for assessing the stability of a dynamic system is the system's response to perturbation. A very stable system will recover rapidly from perturbation; the attractor is strong and tends to "capture" all trajectories in the state space quickly. Less stable systems may also recover from perturbation (i.e., the attractors will be preserved), but their recovery is slower (Schöner & Kelso, 1988). (These predictions hold only for perturbations within a certain range; perturbations exceeding these critical values disrupt the system and may shift it into new attractor regimes.)

In this section, we report on experimentally induced perturbations to treadmill stepping. The first perturbation was the speed change introduced after the first 5 sec of the trial. How rapidly would infants adjust their steps to maintain alternation? Inspection of the foot displacement plots showed that, if the infant was producing alternating steps when the perturbation was introduced, adjustment was immediate and rapid. Figure 21 shows exemplar plots of two infants (CH and BH) at two ages (3 and 7 months) at comparable treadmill speeds at both ages. In some trials, a parallel swing or stance followed the perturbation, but the infant began alternating in the next phase (CH at 3 months, BH at 7 months). In other trials, alternation was not disrupted at all (BH at 3 months, CH at 7 months). CH at 7 months illustrates especially clearly the immediate shorter cycles in response to the treadmill. Corrections seem equally likely at both ages, but it was far less likely for younger infants to be stepping when the perturbation was introduced, which made it difficult to quantify their data.

The second perturbation technique was to disrupt the expected regular alternation of the legs by providing asymmetrical dynamic information to each leg by means of a split-belt treadmill. Recall that the treadmill was constructed with two parallel belts whose speed could be independently controlled. In trials 10 and 11, we adjusted the speed of each belt so that one leg was driven at twice the speed of the other belt. An earlier study showed that, by 7 months, infants precisely adjusted the initiation of the step cycle in each leg to maintain a .5 phase relation; that is, they began the step sooner with the leg on the slow belt and later with the leg on the fast belt to maintain smooth alternation (Thelen, Ulrich, & Niles, 1987). The out-of-phase bilateral alternation pattern thus acted as a stable cyclic attractor: an organization the system "wants" to maintain even when perturbed.

The treadmill trials in the longitudinal study showed that the ability to adjust to the treadmill paralleled general improving performance. In Figure 22, we report the mean number of strictly alternating steps for the two split-belt trials (one trial with the right leg on the fast belt and the left on the slow belt and the second trial with the speeds reversed). In general, performance improved in a linear manner, especially for the infants in the "early" group.

More revealing is the analysis of step cycle durations and relative phase lags performed for the four profiled infants. Recall that the cycle duration is a sensitive reflection of the belt speed: faster speeds lead to shorter cycles. In Figure 23, we compare, for each month, the infants' cycle times on the split belt (one leg moving fast and the other moving slowly) to their cycle times on the belts moving together at similar fast and slow speeds. Across all ages, and with few exceptions, the order of magnitude of cycle durations was as expected; that is, the slow tied-belt condition produced the longest cycle durations, the fast tied-belt produced the shortest durations, and the split-belt conditions resulted in durations that were intermediate to the means for the tied-belt conditions. Note again the striking speedup across all conditions in month 5. Phase lags approached the mature .5 of the cycle with increasing age (Fig. 24). This was true in general for the split- and tied-belt conditions, though the slow split-belt condition, after approaching .5 through month 5, showed a reduction in phase lag in months 6–8.

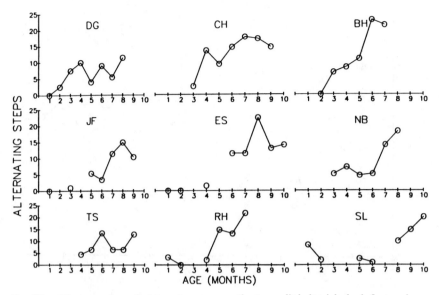

Fig. 22.—Mean number of alternating steps on the two split-belt trials, by infant and age

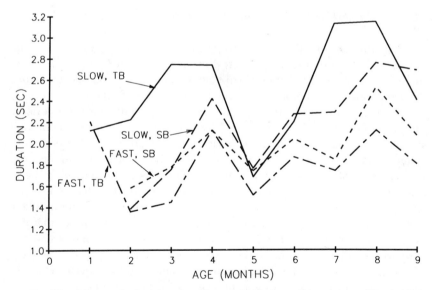

Fig. 23.—Mean cycle durations in split- and tied-belt conditions by age. The tied-belt trials were equated with the fast and slow split-belt speeds. (Slow, TB: foot on the tied belt at slow speed; slow, SB: foot on the slow belt of the split belt; fast, SB: foot on the tied belt at fast speed; fast, TB: foot on the fast belt of the split belt.)

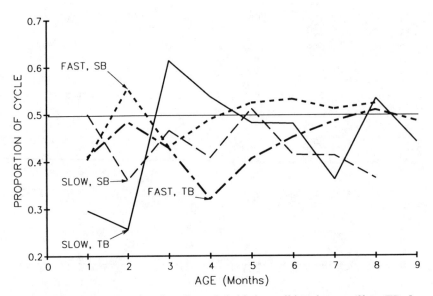

Fig. 24.—Mean phase lags in split- and tied-belt conditions by age. (Slow, TB: foot on the tied belt at slow speed; slow, SB: foot on the slow belt of the split belt; fast, SB: foot on the tied belt at fast speed; fast, TB: foot on the fast belt of the split belt.)

CONTROL PARAMETERS FOR TREADMILL STEPPING

Examination of the dynamic trajectory of treadmill stepping over the first year showed that, although a few infants performed some steps from month 1, all showed a rather abrupt phase shift when their performance (by all measures) dramatically improved. According to dynamic theory, such a phase shift may result when scalar changes in one or more components of the system allow the system to reorganize into new forms. Treadmill stepping requires the cooperative interaction of many components or subsystems, each of which has its own developmental course. In this section, we ask which of several possible contributing components may be acting as the control parameters or the elements engendering the phase shift into stable treadmill stepping. We investigate four concurrently changing likely contributors to treadmill stepping: (1) overall rate of general motor maturation, as indexed by the Bayley scales; (2) developmental changes in the proportion and composition of the legs, as indexed by anthropometric measurements; (3) overall changes in arousal or mood; and (4) specific changes in the predominant postures and movements of the legs that may indicate relative strength of the muscles of the legs. We exploit the naturally occurring differences in stepping onset to ask which of these variables might shift the system.

TABLE 3

CUMULATIVE NUMBER OF BAYLEY ITEMS PASSED BY EACH INFANT AT SUCCESSIVE AGES,
MEAN MONTHLY RATE OF PASSING, AND AGE AT ONSET OF WALKING

INFANT	AGE (Months)										ITEMS PER MONTH	ONSET OF WALKING[a]
	1	2	3	4	5	6	7	8	9	10		
DG......	3	7[b]	8	9	10	13	17	20			2.5	10
CH......	5	6[b]	7	11	15	16	19	20	23		2.6	9
BH......	2	5[b]	8	8	12	15	16				2.3	9
TS	1	5	8[b]	10	12	12	12	16	17		1.9	11
ES.......	2	6	8[b]	10	10	16	16	19	21	23	2.3	13
JF........	4	5	7	8[b]	11	12	17	18	22		2.4	10
RH......	3	3	6	9[b]	12	15	17	19			2.4	10
NB	5	7	7	8	10[b]	14	14				2.0	11.5
SL.......	2	5	7	8	9	11[b]	14	15	18	20	2.0	16

[a] Age in months by parental report.

[b] Indicates the month before onset of steady rise in number of alternate steps taken by the child.

Motor Maturation

One possible control parameter for treadmill stepping may be general motor maturation. That is, the same combination of neurological and morphological factors that contribute to achieving conventional motor milestones, such as rolling over and sitting, may also be acting to shift the infant into improved treadmill stepping. The prediction would be, then, that infants would show some positive relation between their overall motor scores on the Bayley tests and the point when they shifted into stable treadmill stepping. Table 3 reports the cumulative number of Bayley Motor Scale items passed by each infant by month, the number of items passed per month, and the age at which the infant took his or her first independent steps, as reported by a parent. Inspection showed that, while no exact number of items passed was associated with the onset of increasing performance of steps, all infants passed eight items either the month previous to or the month of their noticeable increase; however, these were not the same eight items for each infant. At the same time, the two infants with the latest onset of stepping had a comparatively slower rate of items passed per month than the three earliest steppers, although infant TS had an equally slow rate and was ranked intermediate in stepping onset. The two late steppers were also more delayed in walking onset than the early steppers, but not much different from the intermediate group. The association between overall motor maturation and treadmill stepping was strongest, therefore, at the extremes of the group. It may be that similar factors may have affected overall motor performance, including treadmill stepping and independent walking. It is possible that measuring motor skill acquisition with more discreet measures than the Bayley Scale or measuring motor control variables such as the ability to control voluntarily the relevant muscle groups would uncover a more specific relation between motor maturation and treadmill stepping than we could with the present instrument.

Body Build Characteristics

The general metric of motor maturation is itself a composite variable of many contributing systems: perhaps it is components of motor maturation that influence stepping behavior. Antigravity tasks such as stepping require sufficient muscle strength to lift the mass of the legs. Work by Thelen and her colleagues showed that body build and composition factors could determine motor performance in infancy: infants who had gained mass (presumably fat or nonmuscle tissue) rapidly during the first few months showed rapid declines of upright stepping (Thelen, Fisher, & Ridley-Johnson, 1984). Are body build characteristics control parameters for treadmill stepping? One likely hypothesis is that relatively fat infants, who would presumably have a higher fat-to-muscle ratio and thus proportionately less muscle to lift a greater mass, would be comparatively delayed in the onset of treadmill stepping. Table 4 reports each infant's rank (among the nine subjects) in six anthropometric variables in the month just before his or her rapidly increasing slope of stepping increase. If body build were a determining factor, early steppers would be comparatively leaner and late steppers comparatively chubbier at that point in time. Spearman correlations between infants' ranks in stepping onset and their ranks in body build variables at the transition month were all nonsignificant: infants who stepped early were no different from their late-stepping counterparts in body build characteristics.

Arousal Level

One of the defining characteristics of infant arousal states is motor activity. Sleepy and drowsy infants move infrequently; fussy and crying infants thrash their limbs and have more muscle tension. But behavioral activation is not reflected just in generalized activity. Thelen, Fisher, Ridley-Johnson, and Griffin (1982) found a positive association between arousal level and number of newborn steps in the first 6 weeks. Increased behavioral arousal is presumably reflected in more energy delivered to the muscles and more motor output, including steps. Indeed, it is likely that some threshold level of muscle activation is necessary to lift the legs against gravity. It is thus possible that treadmill stepping was a by-product of the infant's mood. If so, we would predict that infants who stepped on the treadmill frequently

TABLE 4

RANKING OBTAINED BY EACH CHILD ON BODY BUILD VARIABLES 1 MONTH BEFORE ONSET OF RISE IN ALTERNATING STEPS

	Age[b]	ANTHROPOMETRIC VARIABLES[a]							
		Height	Weight	Ponderal Index	Thigh Circumference	Skin Fold	Michelin Index[c]	Leg Length	Leg/ Trunk
DG	2	3	5	4	4	1	6	4	3
CH	2	4	6	7	7	9	7	5	4
BH	2	7	7	3	6	5	2	8	5
TS	3	3.5	3	3	7	5	2	7	9
ES	3	7	5	2	5	2	3	8	7
JF	4	1	1	8	4	2	9	2	5
RH	4	5.5	9	9	9	7	5	1	1
NB	5	7	4	2	3	1	5	6	6
SL	6	8.5	8	5	3	2	4	7.5	3

[a] Higher rank indicates larger values; i.e., infant is longer, heavier, etc.
[b] Age (in months) when the anthropometric measurements were made.
[c] The Michelin Index is the Ponderal Index divided by the thigh circumference (the index is named for the "Michelin" man).

MONOGRAPHS

TABLE 5

MEAN AROUSAL SCORES OBTAINED OVER TEST TRIALS BY EACH INFANT

	AGE (Months)									
INFANT	1	2	3	4	5	6	7	8	9	10
DG	2.91	2.00	2.00	2.00	2.36	2.00	2.00	2.00		
CH	3.33	2.00	2.09	2.18	2.00	2.36	2.00	2.00	3.27	
BH	4.33	2.00	2.18	2.27	2.00	2.00	2.00			
TS	3.33	3.60	2.56	2.09	2.00	2.00	2.00	2.00	2.36	
ES.		2.00	2.36	2.45		2.00	2.00	2.09	2.36	2.64
JF	2.00	2.00	2.00	2.33	2.09	2.00	2.00	2.00	2.00	
RH		2.82	2.86	2.00	2.00	2.00	2.91			
NB	2.67	3.17	2.00	2.00	2.00	2.10	2.09			
SL.	3.00	2.09	2.33	2.56	2.00	2.00	2.00	3.40	2.00	2.09

would be consistently more aroused than those who stepped infrequently. Table 5 shows infants' mean arousal scores for each month. Arousal level was not associated with step performance: correlation between number of steps and mean arousal score for the better day of stepping was low, $r = -.220$, and its direction was opposite that predicted. Infants showed a U-shaped function of arousal level; for most, arousal decreased after the first few months of testing, and several showed an increase again in the last month or two of testing, presumably reflecting their increased impatience with the task. Newborns frequently tolerate this paradigm less well than older babies. They seem to dislike having their clothes off and being held up, which results in increased fussiness. By the time they are pulling to stand on their own, they seem to develop a notion of what the support surface should feel like. The moving belt, therefore, seems strange, and they again become less willing to tolerate it. Indeed, the trend for treadmill steps was in the opposite direction than for newborn steps—increased treadmill steps were associated with lower behavioral arousal. Thus, treadmill stepping required more than just generalized arousal.

Leg Posture

Two characteristics of upright locomotion led us to believe that leg posture (reflecting both neural maturation and muscle characteristics) might index a control parameter for the onset of treadmill stepping.

First, there has been considerable debate over the role of foot contact in the maturity of stepping and walking. In mature locomotion, the heel of the swing leg makes the first contact with the ground. Newborn infants and new walkers contact the surface first with the flat foot or the toe, and new walkers acquire heel strike only after several months. Forssberg (1985) has

76

TABLE 6

LEG POSTURE VARIABLES SHOWING MODERATE CORRELATIONS WITH TOTAL NUMBER OF TREADMILL STEPS AND THEIR INTERCORRELATIONS

	1	2	3	4	5	6	7	8	NUMBER OF STEPS
Legs highly flexed:									
1. When not stepping...		.38	.19	.29	.04	−.33	.05	−.29	−.50
2. In swing phase			.23	.12	.09	−.35	.02	−.18	−.26
Foot orientation inward:									
3. In swing...........				.68	.01	−.13	−.05	−.18	−.26
4. In baseline.........					−.09	−.14	.01	−.30	−.29
Foot contact at touchdown:									
5. On toes............						−.75	.52	−.53	−.46
6. Flat							−.33	.48	.45
Foot contact in stance:									
7. On toes............								−.86	−.43
8. Flat									.60

proposed that the appearance of heel strike results from the maturation of a new central pattern generator that replaces the phylogenetically older flat-footed pattern seen early in development. Thelen and her colleagues (Thelen, 1984; Thelen, Ulrich, & Jensen, 1989) believe that heel strike is an emergent pattern, one that becomes possible only as infants develop strength and control in the stance leg. Thus, the question of the foot patterns in treadmill stepping is important: perhaps one pattern of foot contact is associated with successful stepping.

Second, in both treadmill stepping and overground locomotion, the movement of the leg forward during the swing phase is largely pendular. That is, when the leg is stretched back by either the mechanical action of the treadmill or the shifting of the center of mass over the forward leg, it stores potential energy in stretched muscles and tendons. When the leg is released, it swings forward like a spring, largely without additional muscle contraction. For this to occur, the leg must possess the right amount of "springiness." A too tightly coiled spring will be hard to stretch back, and one coiled too loosely will not develop the required stretch. Thus, whether the leg is flexed or flaccid when both moving and stationary may affect success on the treadmill.

Eight of the 17 leg-posture variables were moderately correlated with the number of alternate steps over all the trials and ages, as shown in Table 6. Poor stepping performance was associated with a high degree of flexion both when belts were moving and the infant was not stepping and during the swing phase when steps were taken; it was also associated with an inward rotation of the foot during baseline (belts-not-moving trials) and during

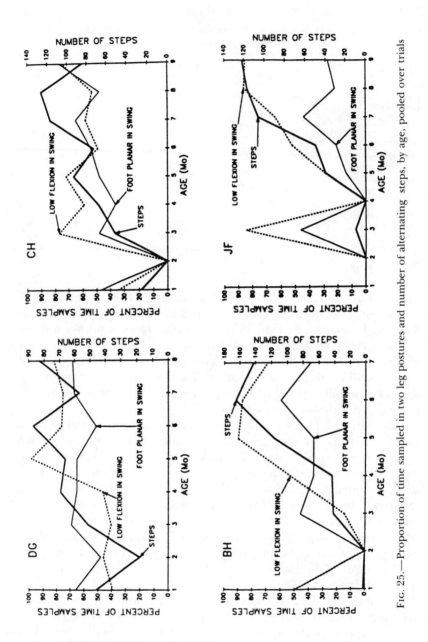

Fig. 25.—Proportion of time sampled in two leg postures and number of alternating steps, by age, pooled over trials

swing. An increase in the number of steps taken was associated with an increase in the frequency of flat-footed contact with the belts both at touch-down from swing and during the stance phase and with a decrease in toe contact. (Heel and lateral foot contact were infrequent.) A flexed-leg position was negatively related to flat-footed contact. Figure 25 illustrates the relation between two leg-posture variables, the proportion of time the leg was in low flexion in swing phase and proportion of time the foot was planar rather than rotated inward or outward and the number of alternating steps for the subsample of four infants.

These results suggest that the orientation of the leg and foot in relation to the moving belts played a crucial role in whether infants responded to the treadmill with alternating steps. When the legs were more tightly flexed, indicated by the predominant leg posture and the toe-first contact with the belt, there may have been insufficient stretch to initiate the stepping pattern. Likewise, when the legs and feet were rotated inward, leading to likely midline crossover while stepping, performance was reduced. It is thus likely that one control parameter for the transition to treadmill stepping is the growth of extensor strength and tone in the leg muscles and the presence of a predominant outward rather than inward leg rotation.

VII. DISCUSSION

To discover how infants and children acquire new behaviors, we must know both the precursor abilities of new skills and the processes by which the old forms are transformed. This study provides a unique window on the coalescence of skill by tracing the course of such a precursor and by suggesting processes of change, explicitly derived from dynamic systems theory.

The treadmill task proved to be especially felicitous for a process-oriented approach for several reasons. First, the stepping elicited by the treadmill is a manifest component of an important fundamental skill, upright locomotion. In mature overground locomotion, the stance leg is stretched back as the weight is shifted onto the descending swing leg. The treadmill mimics this dynamic stretch of the leg and releases an alternating pattern of limb coordination. In infants, the simple mechanical action of the treadmill substitutes for the necessary but immature ability to land on the swing leg and support the weight while shifting the center of gravity over the leg. Thus, there is obvious continuity between treadmill stepping and its mature counterpart.

The behavior is normally cryptic, however, and infants perform it only when the experimenter provides constituents that infants cannot provide themselves. Similarly, neurophysiologists have used treadmills to induce stepping in animals who have been experimentally lesioned and are unable to walk independently. For example, decerebrate and spinalized cats lack balance and the ability to support their weight. However, like infants, they will perform coordinated and responsive stepping when they are supported by the experimenter and their movements are facilitated by the treadmill (for a review, see Grillner, 1980). While we do not know the precise neuro-motor mechanism involved, the addition of the treadmill completes a perception-action loop and elicits complex patterns normally seen only in more mature or intact organisms.

In addition, treadmill stepping differs from many tests of infant abilities in that it requires only minimal cooperation from the infant; indeed, as happened with the older infants in the study, conscious engagement with the task only seemed to impair performance. Therefore, we can be more confident than usual that the task had similar "meaning" to the system from month 1 until well into the last quarter of the first year, allowing us to see changes in the behavior under similar stimulus conditions over a long time span.

TREADMILL STEPPING: A DYNAMIC VIEW

There are a number of ways in which this study supports a dynamic systems view of development and, in particular, the view that upright locomotion emerges from the self-organization of multiple cooperating elements rather than as a result of a preexisting neural code specifying the outcome. Specifically, the treadmill task confirms that behavior is multiply determined and task assembled, that developing systems prefer certain more or less stable attractor states, that the stability of attractors changes as component parameters are scaled, and that systems lose stability at points of transition.

Behavior is multiply determined and task assembled.—That infants can perform treadmill stepping in the first months of life yet do not walk alone until nearly a year later is strong presumptive evidence that even a fundamental skill such as locomotion is multiply determined and that the component elements do not mature synchronously.

Several infants produced speed-responsive treadmill steps as early as 1 month of age. This suggests that they are sensitive to, and organized by, the context or task. The basis for detecting motion (or position) of the legs and using that information to coordinate the movement between the legs is available in the neural anatomy very early in life, perhaps at or near birth. Thelen, Skala, and Kelso (1987) found a similar sensitivity to movement context in another patterned leg movement, spontaneous supine kicking. When they added small weights to one leg, 6-week-old infants responded by changing the speed and amplitude not only of the weighted leg but of the opposite leg as well. As with the treadmill, these infants used their two legs as a responsive synergy. Thus, although these early leg movements are likely not to be intentional, they are highly adaptable and more than just reflexive movements, in any traditional sense of a reflex.

Precocious and context-dependent pattern generation in the limbs early in ontogeny is characteristic of many vertebrate species, including the frog, chick, mouse, rat, and cat. But just as infants require the mechanical action of the treadmill to produce the complex patterned stepping movements, early limb coordination in these other species often also needs facilitative contexts. For example, Stehouwer and Farel (1984) were able to accelerate the onset of hind-limb locomotor activity in frog tadpoles by providing them with a textured surface. Young chicks showed coordinated wing flapping a week earlier than the normal appearance of the behavior when Provine (1981) simply dropped the birds. Newborn rats rarely show coordinated hind-limb activity yet swam with coordinated interlimb movements when submerged (Bekoff & Trainer, 1979). Fentress (1978) elicited precocious and quite complex grooming movements in infant mice by providing them with postural support. And newborn kittens, who similarly will not normally bear their weight and step, would indeed locomote under suitable behavioral conditions, specifically, when they were close to the mother and returning to her side. When isolated from their mothers, the kittens simply extended their hind legs and did not lift their bellies off the ground, owing, presumably, to the interfering effects of their distress and behavioral arousal (Bradley & Smith, 1988).

In each of these cases, as with the human infants, the intrinsic pattern-generating ability of the neuromuscular system must be "completed" by a facilitative environmental context. Often young animals lack sufficient postural support mechanisms to exhibit their underlying coordinative abilities, but functional movement also requires an adequate support surface, surrounding medium, and motivational and emotional states. The functional movement is thus no more contained in the anatomy or in an abstract pattern-generating ability as in the other elements that are essential to the expression of the behavior. None alone is sufficient; all together are necessary.

The elements contributing to the behavior of mature animals—the nature of the support surface and the physical surrounds, the perceptual field, the social and emotional environment—seem so obvious that they can be taken for granted. In infant animals, however, these elements mature asynchronously, and their contributions become clearer. Postural abilities can be distinguished from pattern generation, steps require a nonslippery foothold, and the desire to reach the mother overcomes weak muscles. What allows the behavior to be performed is the interaction of the system components in a particular task context. These components are all in place in adults but are supplied by the experimenter in these infant examples. Thus, human infants can produce step-like patterns at 1 month, and newborn mice can produce grooming-like movements when supported; but the babies are not walking, and the mice are not grooming. Only when the other elements—sometimes seemingly trivial and nonobvious—are also ready can the functional behavior be said to develop. Treadmill stepping may be a manifestation of an important component of walking, but it is not the "essense" of walking or even a privileged component. It indexes only one of many essential elements.

Developing organisms assume more-or-less stable attractor states.—When the developing animal is in a particular context, certain behavioral configurations are preferred over other possible assemblies of the cooperating elements. These attractor states are determined by the morphology of the system, including the system's ontogenetic history and its energy status. We suggest here that, when human infants who are awake and not distressed are placed on the treadmill, the regular alternating step configuration of their legs becomes such an attractor state at some point in the infants' first few months of life. As defined by dynamic systems theory, this pattern was stable over a wide range of input conditions, and the system tended to return to this state when perturbed. We saw that several movement coordinations were possible on the treadmill but that alternating stepping was increasingly preferred.

The measures of alternation between the legs served as a collective variable, that is, the parameter that captures the cooperative interactions of the participating elements. We traced these measures over time in individual developmental trajectories. Despite considerable individual differences in the ages when treadmill stepping became well established, all the infants showed improvements in the number of alternating treadmill steps and their responsiveness to speed and in the between-limb coordination of those steps. Alternation as a response to the treadmill became increasingly attractive with age but was possible, although less likely, even in the first few months of life. A number of measures converged to indicate that the alternating pattern acted as a dynamic attractor for leg movements on the treadmill. Recall that, in dynamic terms, attractor stability is gauged by reduced variability and by rapid recovery from perturbation.

As infants performed more alternating steps on the treadmill, their behavior also became less variable. We saw this in the increasing correspondence between their performance on the 2 days of testing. What was remarkable was that reliability of stepping was performance and not age related, as the late steppers continued to show unstable between-day correspondence until their overall step rate improved. Once alternating stepping was a frequent response to the treadmill, infants were also consistent, suggesting that the state was increasingly preferred.

The more precise measure of interlimb coordination, the actual phase lag calculated for each step, also indicated a more tightly coupled alternation pattern as the number of steps increased. Early steps varied considerably from the mature value of a 50% lag between step initiations, but these values decreased, again as a function of overall ease of response.

That phasing variability seemed to be reduced at certain speeds also suggested that the system was most "comfortable" at particular driving frequencies. (This hypothesis could not be systematically tested here because the sample was too small and the speed metric was not scaled in both directions.) Nonetheless, humans and other animals do perform natural rhythmic limb movements at individually preferred frequencies, which may reflect energy minima for the particular task at hand (Kelso & Scholz, 1985; Kugler & Turvey, 1987; McMahon, 1984). In particular, faster rather than slower speeds appeared to facilitate the coupling of the legs in the infant treadmill situation. Possibly, the faster speed imposed more stringent demands on the timing of the legs if alternation was to be successful even when infants did not have to remain balanced on one leg because the experimenter supported their weight. The puzzling finding of a shorter overall duration at month 5 remains unexplained but warrants further investigation.

Bimanual rhythmic finger and hand movements in adult humans and gait patterns in quadrupeds show qualitative phase shifts in coordination when the actions are intentionally speeded up. A well-known example is the change from walk to trot to gallop in horses, where the distinctive footfall patterns emerge as the animal pumps more energy into its locomotion. Human subjects asked to flex and extend their fingers or wrists at increasing speeds always shifted from an out-of-phase to an in-phase mode of coordination at a characteristic frequency (Kelso & Scholz, 1985). Infants in this study continued to alternate well, however, at all the speeds tested. No particular speed led the pattern to disintegrate or change into another coordinative mode. Perhaps the maximum speed used in the present study was insufficient to elicit such a phase shift, but, since we could not drive our treadmill belts faster, we did not test this possibility.

Perhaps the most dramatic evidence that alternation becomes a preferred behavior is that, as stepping increases, alternation as a coordinative mode comes to dominate the repertoire and the other coordination patterns, theoretically equally likely, drop out. Given any number of possible states, the system in the treadmill task had an increasing probability of residing in only one of them as the infant got older. That this state was selected from a wider universe of possible states confirms that the behavior was a preferred but not obligatory assembly of the contributing components.

Treadmill steps also became more resistant to perturbations as they became the more preferred response. Several perturbations were incorporated into this experiment. In one sense, the changes in speed itself may be considered a perturbation of a steady-state performance. That infants improved in their ability to adjust their step durations to the speed scalar with age suggests that alternation became a more stable response to the conditions of the belt. We increased the speed during a trial, after a 5-sec interval at the previous speed. This slight increase in belt speed, which was similar to a short tug on the system, did not seem to interrupt stepping at any age (although younger infants were relatively unlikely to be continuously stepping when we introduced the speed change). Again, once stepping was established, alternation was a strongly preferred mode.

The attractive nature of alternation was best seen, however, in the split-belt trials where each leg was given highly discrepant information. Overall performance on the split-belt trials improved with age in a manner similar to performance on the tied-belt trials. When we looked at actual cycle durations and phase lags on successful treadmill steps, we saw the expected compensation in the two legs for the speed of the opposite leg at all ages. Infants were able to alternate, and, when they did, they used the information from the movement of one leg to create an adaptive bilateral synergy.

Taken as a whole, these data suggest that an intrinsic ability to step on the moving treadmill in a responsive way is present early in life but that the stability of the attractor improves with age and at different rates in individual infants. Assuming that the behavior requires the cooperation of many subsystems, it appears as though asymptotic performance is limited by other contributing elements than the fundamental perception-action loop. We next consider the nature of the transition into well-developed treadmill stepping and the possible control parameters pushing the system into the new forms.

Using transitions to uncover processes of change.—Dynamic principles make a number of powerful predictions about how complex systems behave at transitions. First, systems shift into new forms because the old forms become unstable. These instabilities should be observable; indeed, they stand as the hallmark of processes of change. Second, scalar changes in one or more components of the system may engender these transitions by disrupting the coherence of the old form and allowing the inherent noise to be amplified into a system-wide reorganization. Thus, components should be identifiable that, when scaled naturally or experimentally, can produce the observed phase shift.

All the infants in this study progressed from a state where they produced no or a few steps and coordination patterns of single, double, and parallel steps to a state where they stepped reliably and well in an alternating pattern. Although performance improved in a linear manner over many months, we could also identify for each infant a month of improvement onset after a shorter or longer plateau, followed by a steep slope of increasing steps. The plateau period itself appeared as a more or less prolonged transition state, demarcated by instabilities and variability in performance and by multiple patterns. Once stepping began to improve, these instabilities were reduced. According to synergetic principles, transitions between stable states are marked by amplification of inherent noise. Because the infants' initial performance was not stable, we could not directly test this prediction; rather, we observed multiple states early on, and that variability was damped out as stepping performance improved.

What then were the control parameters that shifted the system into the phase of steadily improving treadmill performance? Here we exploited the considerable individual differences among the infants' onset phase for clues to what may have changed behavior forms, assuming that the direct sensori-motor pathways mediating the adaptive response to the treadmill were already in place.

Although general body build characteristics have been implicated in other antigravity tasks, they did not have any association with treadmill stepping. Early and later steppers were equally lean or chubby. Perhaps the mechanical action of the treadmill imparted sufficient energy to the leg to overcome any deficiencies in muscle strength compared to leg mass. Likewise, predominant arousal level or mood did not distinguish between infants and was unlikely to be a factor in stepping improvement.

The most likely candidates for control parameters were whatever combination of neurological and musculoskeletal factors that maintained flexor rather than extensor dominance in the legs and an outward and planar rather than inward rotation of the legs and feet. If, as we have suggested, treadmill stepping is elicited either by the dynamic stretch of the leg backward on the moving belt or by some threshold value of the hip position relative to the trunk, then the spring tension on the leg must permit this leg stretch. Newborn infants have a characteristic flexor dominance in their limbs that only gradually wanes during the first 6 months. This flexor preference is seen in both posture (Gesell, 1939) and movement (Thelen, 1985). For example, in kicking and stepping movements in the first few months, the flexion phase appears to be initiated by active muscle contraction while leg extension is a result of largely passive elastic and inertial forces (Schneider, Zernicke, Ulrich, Jensen, & Thelen, 1990; Thelen & Fisher, 1983).

Why are newborn and young infants flexor dominant? First, although it seems intuitively easy to distinguish a "floppy" infant from one with stiff or tight muscles, the concept of muscle tone is not easily operationalized or well understood. This is because the quality of tone comprises both active motorneuron recruitment and more passive, elastic qualities of the muscles. Both of these are likely to contribute to the tone of the young infant. Maekawa and Ochiai (1975) measured the EMG response to passive extension of the limbs (the active portion) and the resting elbow and knee joint angles (the passive component) of infants from birth to 8 days of age. During the first 48 hours, the newborns maintained their limbs in a tightly flexed posture and showed strong recoil when extended. However, the EMG activity in the recoil was comparatively low. After 48 hours, the flexed posture decreased somewhat, as did the recoil response, but the EMG reaction to passive resistance became greater. This suggested that, in the first 2 days, a position-induced bias of the muscles, presumably from the long confinement in the tightly packed intrauterine space, strongly influenced the flexor posture and movement. However, after a few days, as infants gained strength, active muscle contraction became increasingly important in the flexor bias. For example, Schulte and Schwenzel (1965) found that flexor muscle groups in newborns showed long-lasting tonic activity but that extensor groups did not, although both showed phasic activity. Thelen (1985) also reported higher EMG bursts in flexor muscles.

At birth, infants are both released from the tight confinement of the uterus and subject to gravity, and it is likely that the gradual relaxation of the flexor bias reflects this experiential change. Breech infants, for instance, who have a more extended posture in utero also have characteristically extended legs as newborns (Maekawa & Ochiai, 1975). Further evidence that the uterine constraints may have biased the muscles comes from studies of premature infants. Heriza (1988) showed that premature infants who spent 6–12 weeks out of utero had a more extended leg posture and a greater range of motion than full-term controls. The transition to extrauterine life is a good example of how nonspecific environmental constraints can affect neuromotor development.

The alternating flexor and extensor influences in posture and movement during the first year suggest asynchronous growth in flexor and extensor muscle groups or in their neurological control (Gesell, 1939; Thelen, 1985). Within this waxing and waning, extensor strength seems to lag behind, as reflected in head and trunk control as well as limb control. Indeed, Sutherland, Olshen, Cooper, & Woo (1980) have suggested that one important element constraining independent walking is the relative weakness of the ankle extensor muscles, which are essential for stabilizing the foot as it bears weight in stance. (If the leg buckles, locomotion is impossible.) New walkers as well as treadmill walkers may contact the ground with their flat feet or their toes because they lack the extensor ability to dorsiflex the foot for heel contact.

If flexor-extensor balance is a control parameter for treadmill stepping, this may explain the association between the onset of treadmill stepping and general motor maturity that was evident especially for the earliest and latest steppers. Recall that control parameters may be entirely nonspecific to the variable of interest. The somewhat delayed onset of extensor strength that retarded treadmill stepping in some infants may also have caused the lags in other motor tasks that indeed require extensor control. For example, many items on the Bayley test measure head and trunk control directly, and others, such as reaching tasks, measure such postural control indirectly because a stable posture is necessary for articulated limb movements. In addition, the walking delay seen in the latest-onset steppers may also reflect this comparative lack of extensor strength. Since treadmill stepping was never practiced directly by these infants outside the twice-monthly laboratory sessions, improvements in treadmill stepping must have been the result of changes in components exercised in other motor contexts as well.

In sum, it is instructive to plot the collective variable, that is, number

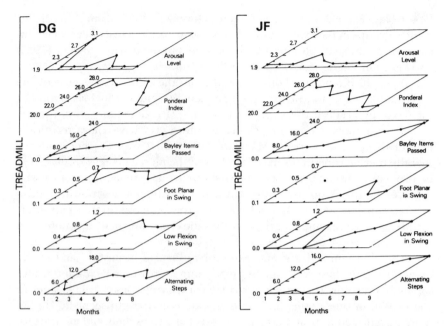

FIG. 26.—The development of alternating treadmill steps depicted as parallel developmental trajectories of related subsystems.

of alternating steps, in parallel with the developmental trajectories of the potential control parameters as the layered system depicted in Figure 7. In Figure 26, we do this for an early and a late stepper. Infant DG's stepping in the first session may have been facilitated by a combination of high arousal, which "energized" the steps, relatively low Ponderal index, and a planar orientation of the foot and leg. In contrast, JF's poorer early step performance may have resulted from his being relatively more chubby combined with low arousal and nonoptimal foot posture. Although without further confirmation the results remain correlational and only suggestive, we have a particularly clear picture of the multiply interacting subsystems and how they change over time. This allows us to identify manipulations of putative control parameters at phase shifts, as we discuss next.

Manipulating putative control parameters.—The evidence from this longitudinal study is suggestive, but certainly not conclusive, that particular neurological and musculoskeletal properties that constitute the quality of limb tone act as a control parameter for the performance of treadmill stepping. The next step in a dynamic process analysis of change would be to manipulate the putative control parameters experimentally to see if the system could be shifted in the predicted direction. For example, Thelen and Fisher

(1982) suggested that rapid weight gain, a nonspecific growth effect, acted to depress newborn stepping by adding too much nonmuscle mass to the legs for the muscles to lift. This was supported by experiments in which Thelen, Fisher, and Ridley-Johnson (1984) artificially changed the effective mass of the legs by either adding growth-appropriate weights or partially submerging the infants in water. Stepping decreased or increased in the predicted directions.

Unfortunately, the relative flexor dominance of the legs is difficult to simulate in an experimental situation and was not tested in the present study. Several nonnormal populations of infants may provide additional confirmation. Ulrich and Ulrich (1990) tested a sample of Down syndrome infants in a longitudinal treadmill design. These infants are characteristically hypotonic and show delayed motor milestones. Results indicated comparatively delayed onset of treadmill stepping as well, suggesting, perhaps, insufficient muscle tension to benefit from the treadmill action. In addition, Davis and Thelen are testing a group of premature infants during their first year. These infants are at risk for motor problems, most commonly spasticity. Preliminary results in this group are remarkable and support the current hypothesis: at 1 month corrected age, premature infants are dramatically better treadmill steppers than full-term infants at the same age. Recall that Heriza (1988) discovered that premature infants with 6 weeks or more experience out of utero were distinctly more extensor dominant in kicking. If these results are confirmed, it may also be that the experience in an extended posture and in gravity also facilitates interaction with the treadmill. Indeed, it is tempting to speculate that a 6-week premature infant tested at 1 month corrected age has had about the same out-of-utero experience as a normal infant does at the time when treadmill stepping performance accelerates. It is important to note, however, that any results with these nonnormal groups must be interpreted carefully. All these conditions have multiple organic and environmental consequences, and the direction of causality is difficult to ascertain.

We expect to learn more about the kinematics and biomechanics of treadmill stepping from our current research program aimed at investigating the intersegmental dynamics of infant movements (Schneider et al., 1990; Ulrich, Jensen, & Thelen, 1991). These techniques and models allow us to examine not only the time and space behavior of the limbs but also the actual apportionment of forces generated during movement. In this way, we can differentiate the contribution of each joint to the movement from the effects of gravity, of active muscle contraction, and of the inertial forces generated by movements of the other segments of the limb. If we are correct in our hypothesis that leg posture is a control parameter, then we would expect to see a shift in the predominant muscle torques at the time of step improvement. That is, poor stepping ability would be associated with the predominant muscle pull on the joint being in the flexor direction, with a lessening of the flexor and increasing influence of extensor torques accompanying better steps. Preliminary results of steps in 7-month-old infants (who are good treadmill steppers) indicate that even at this age the swing phase of the treadmill step is still more flexor dominant than the swing of a treadmill step in adults. That is, adults appear to use more passive and elastic forces to swing the leg forward. More study is clearly indicated.

TREADMILL STEPPING AND LEARNING TO WALK

This study supports our earlier hypothesis about the development of locomotion as an emergent rather than a prescribed process (Thelen, 1984, 1986a; Thelen, Kelso, & Fogel, 1987; Thelen, Ulrich, & Jensen, 1989). Just as treadmill stepping itself requires not only intrinsic pattern generation but also facilitative musculoskeletal conditions and a supportive biomechanical context, so independent locomotion develops as a confluence of equally necessary factors. We suggested previously that the ability to generate locomotor-like patterns in the legs was evident in infant supine kicking. The ontogenetic link between early pattern generation and actual locomotion is strengthened, however, with the discovery of treadmill stepping as continuously available from early infancy.

Moreover, the fact that treadmill stepping is possible but that the behavior is never performed without the treadmill helps explain why step-like pattern generation is in place in the early months but why infants do not walk until about 1 year of age. What seems to be crucial is the stretching of the stance leg backward as the weight shifts forward onto the descending swing leg. Without the treadmill and the full postural support of the experimenter, this would require that infants support their weight fully on one leg and maintain their balance while swinging one leg forward. The former requires a nearly rigid leg with strong extensor support; the latter requires considerable skill to correct for the inherent instability of single leg support (Sutherland et al., 1980). Both of these appear to be rather late-maturing abilities in contrast to the other elements contributing to locomotion, such as motivation, voluntary control of the legs, spatial mapping of the environment, or the ability to use information from the visual flow field to guide movement. The fact that these other elements coalesce earlier in crawling or creeping is further evidence that the task of upright, bipedal locomotion is constrained primarily by strength and balance factors.

One tenet of a dynamic approach is that, as systems develop, they change in a nonlinear fashion; thus, what moves the system into new forms at one time may not be the control parameter at another (Thelen, 1988).

The likely control parameter that shifted infants into the phase of treadmill stepping was nonspecific changes in the neuromuscular apparatus, not practice in the behavior itself. The later transition from not walking to walking alone may have different control parameters, however. In particular, we could speculate that specific experience with standing, supporting their weight, and "cruising" allows infants to explore the state space of their own dynamic balance, the nature of the support surfaces, and their movement-related visual environment. Each movement for balance or traveling is associated not only with its functional consequences but also with the multimodal feedback from the visual, vestibular, tactile, and proprioceptive receptors that are concurrently activated. There is compelling anatomical evidence that the brain encodes these perception-action correlates in many parallel and overlapping circuits, allowing certain associative networks to become strengthened with experience (Edelman, 1987; Kuperstein, 1988). As movements are repeated over and over in similar, but not identical, ways, maps are built of these associations, and functional actions are thus "selected" from the wider universe of possibilities. These preferred behaviors act again as dynamic attractors, increasing in stability but remaining functionally flexible, much like alternating stepping became molded as a functional attractor during the first half of the first year (Thelen, 1990a).

Through repeated practice of the components of locomotion, then, the system self-assembles into the new form of independent upright walking that then itself becomes more stable. In dynamic terminology, this new attractor "captures" other trajectories and becomes the highly preferred mode of locomotion. For example, a new walker may resort to crawling when the motivation to move quickly is high or the support surface is unfamiliar or too slippery. In a few weeks or months, the infant abandons crawling altogether and chooses to walk despite a variety of initial conditions and any number of potential and real perturbations to the path. Likewise, new walkers are easily perturbed by manipulations in the visual field, but, with practice, the attractor becomes so stable that compensations are rapid and effective (Gibson & Schmuckler, 1989).

BEYOND TREADMILL STEPPING: IMPLICATIONS FOR DEVELOPMENTAL ANALYSIS

The treadmill revealed a cryptic component of a fundamental perceptual-motor skill that was best understood by a dynamic systems analysis. The fundamental assumptions of a dynamic approach—namely, self-organization from multiple components, attractor states rather than mental schemata or neural programs, and the transition to new forms from loss of stability—apply equally well, we believe, to developmental processes in many other domains (see, e.g., Fogel & Thelen, 1987; Thelen, 1989, 1990b, in press).

On a theoretical level, a dynamic systems approach is particularly successful in explaining a number of long-standing puzzles of ontogeny, such as the origins of new forms, the stage-like coherence of behavior from underlying continuous processes, and the phenomena of décalage. Because new forms are emergent and entirely context dependent, there is no privileged element to be revealed either in some abstract construct or in the genetic code. The importance of the contextual milieu gives the environment a causally equal, but again not privileged, role in performance. Social and cultural factors are as deterministic as organic factors. Because the participating elements themselves develop asynchronously, bits and pieces of more mature behavior may be revealed by the proper contextual conditions; conversely, organisms may revert to less mature behavior when conditions disrupt the ongoing stable states. Thus, while the preferred behavioral configurations give the appearance of stages of development—under normal and usual conditions the system assumes certain probable configurations—these stages are statistical descriptions rather than prescriptively encoded programs. Finally, this approach suggests that change is a result of the loss of system coherence through amplification of inherent noise and the autonomous seeking of new solutions. Continuous changes in one or a few organic or environmental components are sufficient to cause wide-ranging reorganization. Developing systems are nonlinear and may respond with large effects to small changes.

The instantiation of these principles in any domain remains, however, entirely an empirical enterprise. They predict nothing about any particular developmental problem in the absence of detailed data. The strategy for collecting such data needs to be different, however, from the conventional cross-sectional group design. Because the approach relies on measures of stability, instability, and transitions, longitudinal design is essential for mapping the developmental trajectory. Second, the individual must be the unit of analysis. We would have completely lost the power to see transitions and to exploit differences if we had looked at treadmill stepping in groups of infants at 1, 3, and 5 months, for instance. We would have treated our variability as noise around some group mean rather than as a source of information about what is stable and what changes. That some infants could perform steps at 1 month and others not until 5 or 6 months tells us about the nature of the mechanisms and processes. The identification of transitions opens up a window to explore process and to identify those variables that move the system into new forms. There is considerable evidence that developmental transitions fulfill dynamic predictions (e.g., Siegler & Jenkins, 1989), and we look forward to future tests of these concepts.

REFERENCES

Bayley, N. (1969). *Bayley Scales of Infant Development.* New York: Psychological Corp.

Bekoff, A., & Trainer, W. (1979). The development of interlimb coordination during swimming in postnatal rats. *Journal of Experimental Biology, 83,* 1–11.

Bertalanffy, L. von. (1968). *General system theory.* New York: Braziller.

Bertenthal, B. I., Campos, J. J., & Barrett, K. C. (1984). Self-produced locomotion: An organizer of emotional, cognitive, and social development in infancy. In R. N. Emde & R. J. Harmon (Eds.), *Continuities and discontinuities in development* (pp. 175–210). New York: Plenum.

Bradley, N. S., & Smith, J. L. (1988). Neuromuscular patterns of stereotypic hindlimb behaviors in the first two postnatal months: 1. Stepping in normal kittens. *Developmental Brain Research, 38,* 37–52.

Brainerd, C. J. (1978). The stage question in cognitive-developmental theory. *Behavioral and Brain Sciences, 1,* 173–182.

Brent, S. B. (1978). Prigogine's model for self-organization in nonequilibrium systems: Its relevance for developmental psychology. *Human Development, 21,* 374–387.

Chapman, M. (1988). *Constructive evolution: Origins and development of Piaget's thought.* New York: Cambridge University Press.

Davies, P. (1988). *The cosmic blueprint: New discoveries in nature's creative ability to order the universe.* New York: Simon & Schuster.

Diamond, A. (1990). Differences between adult and infant cognition: Is the crucial variable presence or absence of language? In L. Weiskrantz (Ed.), *Thought without language* (pp. 337–370). Oxford: Clarendon.

Edelman, G. M. (1987). *Neural Darwinism.* New York: Basic.

Fentress, J. C. (1978). *Mus muscicus:* The developmental orchestration of selected movement patterns in mice. In G. M. Burghardt & M. Bekoff (Eds.), *The development of behavior: Comparative and evolutionary aspects* (pp. 321–342). New York: Garland STPM.

Fiorentino, M. R. (1981). *A basis for sensorimotor development—normal and abnormal: The influence of primitive, postural reflexes on the development and distribution of tone.* Springfield, IL: Thomas.

Fischer, K. W. (1987). Relations between brain and cognitive development. *Child Development, 58,* 623–632.

Fischer, K. W., & Bidell, T. R. (in press). Constraining nativist inferences about cognitive capacities. In S. Carey & R. Gelman (Eds.), *Constraints on knowledge in cognitive development.* Hillsdale, NJ: Erlbaum.

Fogel, A., & Thelen, E. (1987). The development of expressive and communicative action in the first year: Reinterpreting the evidence from a dynamic systems perspective. *Developmental Psychology, 23,* 747–761.

Forssberg, H. (1985). Ontogeny of human locomotor control: 1. Infant stepping, supported locomotion, and transition to independent locomotion. *Experimental Brain Research,* **57,** 480–493.

Forssberg, H., & Wallberg, H. (1980). Infant locomotion: A preliminary movement and electromyographic study. In K. Berg & B. Eriksson (Eds.), *Children and exercise* (Vol. **9,** pp. 32–40). Baltimore, MD: University Park Press.

Gelman, R., & Baillargeon, R. (1983). A review of some Piagetian concepts. In J. H. Flavell & E. M. Markman (Eds.), P. H. Mussen (Series Ed.), *Handbook of child psychology: Vol. 3. Cognitive development* (pp. 167–230). New York: Wiley.

Gesell, A. (1939). Reciprocal interweaving in neuromotor development. *Journal of Comparative Neurology,* **70,** 161–180.

Gibson, E. J., & Schmuckler, M. A. (1989). Going somewhere: An ecological and experimental approach to development of mobility. *Ecological Psychology,* **1,** 3–25.

Glass, L., & Mackey, M. C. (1988). *From clocks to chaos: The rhythms of life.* Princeton, NJ: Princeton University Press.

Gleick, J. (1987). *Chaos: Making a new science.* New York: Viking.

Goel, N. S., Doggenweiler, C. F., & Thompson, R. L. (1986). Simulation of cellular compaction and internalization in mammalian embryo development as driven by minimization of surface energy. *Bulletin of Mathematical Biology,* **48,** 167–187.

Goldman-Rakic, P. S. (1987). Development of cortical circuitry and cognitive function. *Child Development,* **58,** 601–622.

Goodwin, B. C., & Trainor, L. E. H. (1985). Tip and whorl morphogenesis in *Acetabularia* by calcium-regulated strain fields. *Journal of Theoretical Biology,* **117,** 79–106.

Grebogi, C., Ott, E., & Yorke, J. A. (1987). Chaos, strange attractors, and fractal basin boundaries in nonlinear dynamics. *Science,* **238,** 632–638.

Greenough, W. T., Black, J. E., & Wallace, C. S. (1987). Experience and brain development. *Child Development,* **58,** 539–559.

Grillner, S. (1980). Control of locomotion in bipeds, tetrapods, and fish. In V. B. Brooks (Ed.), *Handbook of physiology: Vol. 3. Motor control* (pp. 1179–1236). Bethesda, MD: American Physiological Society.

Grossberg, S., & Kuperstein, M. (1986). *Neural dynamics of adaptive sensory motor control: Ballistic eye movements.* Amsterdam: Elsevier.

Haken, H. (1977). *Synergetics: An introduction.* Heidelberg: Springer.

Haken, H. (1983). *Synergetics, an introduction: Non-equilibrium phase transitions and self-organization in physics, chemistry, and biology* (3d ed.). Berlin: Springer.

Haken, H., Kelso, J. A. S., & Bunz, H. (1985). A theoretical model of phase transitions in biological systems. *Biological Cybernetics,* **39,** 139–156.

Heriza, C. B. (1988). Comparison of leg movements in preterm infants at term with healthy full-term infants. *Physical Therapy,* **68,** 1687–1693.

Hofsten, C. von. (1989). Motor development as the development of systems. *Developmental Psychology,* **25,** 950–953.

Inman, V. T., Ralston, H. J., & Todd, F. (1981). *Human walking.* Baltimore: Williams & Wilkins.

Kelso, J. A. S. (1984). Phase transitions and critical behavior in human bimanual coordination. *American Journal of Physiology,* **15,** A1000–A1004.

Kelso, J. A. S., Mandell, A. J., & Shlesinger, M. F. (Eds.). (1988). *Dynamic patterns in complex systems.* Singapore: World Scientific.

Kelso, J. A. S., & Scholz, J. P. (1985). Cooperative phenomena in biological motion. In H. Haken (Ed.), *Complex systems: Operational approaches in neurobiology, physics, and computers* (pp. 124–149). Heidelberg: Springer.

Kelso, J. A. S., Scholz, J. P., & Schöner, G. (1986). Non-equilibrium phase transitions in coordinated biological motion: Critical fluctuations. *Physics Letters A,* **118,** 279–284.

Kitchener, R. F. (1982). Holism and the organismic model in developmental psychology. *Human Development*, **25**, 233–249.

Kugler, P., Kelso, J. A. S., & Turvey, M. T. (1982). On the control and coordination of naturally developing systems. In J. A. S. Kelso & J. E. Clark (Eds.), *The development of movement control and coordination* (pp. 5–78). New York: Wiley.

Kugler, P. N., & Turvey, M. T. (1987). *Information, natural law, and the self-assembly of rhythmic movement.* Hillsdale, NJ: Erlbaum.

Kuhn, D., & Phelps, E. (1982). The development of problem-solving strategies. In H. Reese & L. Lipsitt (Eds.), *Advances in child development and behavior* (Vol. **17**, pp. 2–44). New York: Academic.

Kuperstein, M. (1988). Neural network model for adaptive hand-eye coordination for single postures. *Science*, **239**, 1308–1311.

Laszlo, E. (1972). *Introduction to systems philosophy.* New York: Harper & Row.

Leonard, C. T., Hirschfeld, H., & Forssberg, H. (1988). Gait acquisition and reflex abnormalities in normal children and children with cerebral palsy. In B. Amblard, A. Bertoz, & F. Clarac (Eds.), *Posture and gait: Development, adaptation, and modulation* (pp. 33–45). Amsterdam: Elsevier Science.

Lerner, R. M. (1978). Nature, nurture, and dynamic interaction. *Human Development*, **21**, 1–20.

Maekawa, K., & Ochiai, Y. (1975). Electromyographic studies on flexor hypertonia of the extremities of newborn infants. *Developmental Medicine and Child Neurology*, **17**, 440–446.

McGraw, M. B. (1945). *The neuromuscular maturation of the human infant.* New York: Columbia University Press.

McMahon, T. A. (1984). *Muscles, reflexes, and locomotion.* Princeton, NJ: Princeton University Press.

Meakin, P. (1986). A new model for biological pattern formation. *Journal of Theoretical Biology*, **118**, 101–113.

Mittenthal, J. E. (1981). The rule of normal neighbors: A hypothesis for morphogenetic pattern regulation. *Developmental Biology*, **88**, 15–26.

Odell, G. M., Oster, G., Alberch, P., & Burnside, B. (1981). The mechanical basis of morphogenesis: 1. Epithelial folding and invagination. *Developmental Biology*, **85**, 446–462.

Oyama, S. (1985). *The ontogeny of information: Developmental systems and evolution.* Cambridge: Cambridge University Press.

Piaget, J. (1985). *The equilibration of cognitive structures: The central problem of intellectual development.* Chicago: University of Chicago Press.

Prigogine, I., & Stengers, I. (1984). *Order out of chaos: Man's new dialogue with nature.* New York: Bantam.

Provine, R. R. (1981). Development of wing-flapping and flight in normal and flap-deprived domestic chicks. *Developmental Psychobiology*, **14**, 279–291.

Rovee-Collier, C. K., & Fagen, J. W. (1981). The retrieval of memory in early infancy. In L. P. Lipsitt & C. K. Rovee-Collier (Eds.), *Advances in infancy research* (Vol. **1**, pp. 225–254). Norwood, NJ: Ablex.

Sameroff, A. J. (1983). Developmental systems: Contexts and evolution. In W. Kessen (Ed.), P. H. Mussen (Series Ed.), *Handbook of child psychology: Vol. 1. History, theory, and methods* (pp. 237–294). New York: Wiley.

Schaap, P. (1986). Regulation of size and pattern in the cellular slime molds. *Differentiation*, **33**, 1–16.

Schmidt, R. C., Carello, C., & Turvey, M. T. (1990). Phase transitions and critical fluctuations in the visual coordination of rhythmic movements between people. *Journal of Experimental Psychology: Human Perception and Performance*, **16**, 227–247.

Schneider, K., Zernicke, R. F., Ulrich, B. D., Jensen, J. L., & Thelen, E. (1990). Understanding movement control in infants through the analysis of limb intersegmental dynamics. *Journal of Motor Behavior,* **22,** 493–520.

Schöner, G., & Kelso, J. A. S. (1988). Dynamic pattern generation in behavioral and neural systems. *Science,* **239,** 1513–1520.

Schulte, F. J., & Schwenzel, W. (1965). Motor control and muscle tone in the newborn period: Electromyographic studies. *Biology of the Neonate,* **8,** 198–215.

Seigler, R. S., & Jenkins, E. A. (1989). *How children discover new strategies.* Hillsdale, NJ.: Erlbaum.

Sekimura, T., & Kobuchi, Y. (1986). A spatial pattern formation model for *Dictyostelium discoideum. Journal of Theoretical Biology,* **122,** 325–338.

Shaw, R. (1984). *The dripping faucet as a model chaotic system.* Santa Cruz, CA: Aerial.

Soll, D. R. (1979). Timers in developing systems. *Science,* **203,** 841–849.

Stehouwer, D. J., & Farel, P. B. (1984). Development of hindlimb locomotor behavior in the frog. *Developmental Psychobiology,* **17,** 217–232.

Sutherland, D. H., Olshen, R., Cooper, L., & Woo, S. L.-Y. (1980). The development of mature gait. *Journal of Bone and Joint Surgery,* **62,** 336–353.

Tapaswi, P. K., & Saha, A. K. (1986). Pattern formation and morphogenesis: A reaction-diffusion model. *Bulletin of Mathematical Biology,* **48,** 213–228.

Thelen, E. (1984). Learning to walk: Ecological demands and phylogenetic constraints. In L. P. Lipsitt (Ed.), *Advances in infancy research* (Vol. **3,** pp. 213–250). Norwood, NJ: Ablex.

Thelen, E. (1985). Developmental origins of motor coordination: Leg movements in human infants. *Developmental Psychobiology,* **18,** 1–22.

Thelen, E. (1986a). Development of coordinated movement: Implications for early human development. In M. G. Wade & H. T. A. Whiting (Eds.), *Motor skills acquisition* (pp. 107–124). Dordrecht: Nijhoff.

Thelen, E. (1986b). Treadmill-elicited stepping in seven-month-old infants. *Child Development,* **57,** 1498–1506.

Thelen, E. (1987). The role of motor development in developmental psychology: A view of the past and an agenda for the future. In N. Eisenberg (Ed.), *Contemporary topics in developmental psychology* (pp. 3–33). New York: Wiley.

Thelen, E. (1988). Dynamical approaches to the development of behavior. In J. A. S. Kelso, A. J. Mandell, & M. F. Shlesinger (Eds.), *Dynamic patterns in complex systems* (pp. 348–369). Singapore: World Scientific.

Thelen, E. (1989). Self-organization in developmental processes: Can systems approaches work? In M. Gunnar & E. Thelen (Eds.), *Systems in development: The Minnesota Symposia in Child Psychology* (Vol. **22,** pp. 77–117). Hillsdale, NJ: Erlbaum.

Thelen, E. (1990a). Coupling perception and action in the development of skill: A dynamic approach. In H. Bloch & B. Bertenthal (Eds.), *Sensory-motor organization and development in infancy and early childhood* (pp. 39–56). Dordrecht: Kluwer Academic.

Thelen, E. (1990b). Dynamical systems and the generation of individual differences. In J. Colombo & J. W. Fagen (Eds.), *Individual differences in infancy: Reliability, stability, and prediction* (pp. 19–43). Hillsdale, NJ: Erlbaum.

Thelen, E. (in press). Motor aspects of emergent speech. In N. Krasnegor (Ed.), *Biobehavioral foundations of language.* Hillsdale, NJ: Erlbaum.

Thelen, E., & Cooke, D. W. (1987). The relationship between newborn stepping and later locomotion: A new interpretation. *Developmental Medicine and Child Neurology,* **29,** 380–393.

Thelen, E., & Fisher, D. M. (1982). Newborn stepping: An explanation for a "disappearing reflex." *Developmental Psychology,* **18,** 760–775.

Thelen, E., & Fisher, D. M. (1983). The organization of spontaneous leg movements in newborn infants. *Journal of Motor Behavior, 15,* 353–377.

Thelen, E., Fisher, D. M., & Ridley-Johnson, R. (1984). The relationship between physical growth and a newborn reflex. *Infant Behavior and Development, 7,* 479–493.

Thelen, E., Fisher, D. M., Ridley-Johnson, R., & Griffin, N. (1982). The effects of body build and arousal on newborn infant stepping. *Developmental Psychobiology, 15,* 447–453.

Thelen, E., Kelso, J. A. S., & Fogel, A. (1987). Self-organizing systems and infant motor development. *Developmental Review, 7,* 39–65.

Thelen, E., Skala, K., & Kelso, J. A. S. (1987). The dynamic nature of early coordination: Evidence from bilateral leg movements in young infants. *Developmental Psychology, 23,* 179–186.

Thelen, E., & Smith, L. S. (in preparation). *A dynamical theory of the development of action, perception and cognition.*

Thelen, E., Ulrich, B. D., & Jensen, J. L. (1989). The developmental origins of locomotion. In M. Woollacott & A. Shumway-Cook (Eds.), *The development of posture and gait across the lifespan* (pp. 25–47). Columbia: University of South Carolina Press.

Thelen, E., Ulrich, B., & Niles, D. (1987). Bilateral coordination in human infants: Stepping on a split-belt treadmill. *Journal of Experimental Psychology: Human Perception and Performance, 13,* 405–410.

Turiel, E., & Davidson, P. (1986). Heterogeneity, inconsistency, and asynchrony in the development of cognitive structures. In I. Levin (Ed.), *Stage and structure: Reopening the debate* (pp. 106–143). Norwood, NJ: Ablex.

Ulrich, B. D., Jensen, J. L., & Thelen, E. (1991). Stability and variation in the development of infant stepping: Implications for control. In A. E. Patla (Ed.), *Adaptability of human gait: Implications for the control of locomotion* (pp. 145–164). Amsterdam: Elsevier Science.

Ulrich, B. D., & Ulrich, D. A. (1990, May). Deviant motor development: Can dynamical systems theory provide an explanation? In J. Whitall (Chair), *Developmental aspects of interlimb coordination.* Symposium conducted at the annual meeting of the North American Society for the Psychology of Sport and Physical Activity, Houston.

van Geert, P. (1991). A dynamic systems model of cognitive and language growth. *Psychological Review, 98,* 3–53.

Vygotsky, L. S. (1962). *Thought and language.* Cambridge, MA: MIT Press.

Weiss, P. A. (1969). The living system: Determinism stratified. In A. Koestler & J. R. Smithies (Eds.), *Beyond reductionism: New perspectives in the life sciences* (pp. 3–55). Boston: Beacon.

Wilt, F. H. (1987). Determination and morphogenesis in the sea urchin embryo. *Development, 100,* 559–575.

Wolff, P. H. (1987). *The development of behavioral states and the expression of emotions in early infancy: New proposals for investigation.* Chicago: University of Chicago Press.

Woollacott, M. H., Shumway-Cook, A., & Williams, H. (1989). The development of posture and balance control in children. In M. Woollacott & A. Shumway-Cook (Eds.), *The development of posture and gait across the lifespan* (pp. 77–96). Columbia: University of South Carolina Press.

Yates, F. E. (1987). *Self-organizing systems: The emergence of order.* New York: Plenum.

Zelazo, P. R. (1976). From reflexive to instrumental behavior. In L. P. Lipsitt (Ed.), *Developmental psychobiology: The significance of infancy* (pp. 87–104). Hillsdale, NJ: Erlbaum.

ACKNOWLEDGMENTS

We dedicate this *Monograph* with respect and affection to the parents, who had to explain to Grandma why those people at the university were attaching little light bulbs to their naked baby and making her walk on a treadmill, and to the infants, who allowed us to do it.

David Niles and Debbie Gorney were instrumental in all aspects of data collection and analysis; we could not have done this without their assistance. We also thank Elizabeth Burrello, Lynn Covitz, Anne Mathews, Mike Schoeny, and Mary Twohy for many hours of dedicated help, Douglas Collier for his interpretation of the neurological literature, and Jody Jensen and Dexter Gormley for many hours of patient work on the graphics. Dan Bullock, Linda Smith, Steve Robertson, and an anonymous reviewer made extremely helpful comments on various stages of the manuscript. We tried to listen.

This study was supported by National Science Foundation grant BNS 85-09793, by National Institutes of Health grant RO1 HD 22830, by a Research Career Development Award and a Research Scientist Development Award to Esther Thelen, and by funds provided by Research and Graduate Development, Indiana University.

Address correspondence to Esther Thelen, Department of Psychology, Indiana University, Bloomington, IN 47405.

HOW ARE NEW BEHAVIORAL FORMS AND FUNCTIONS
INTRODUCED DURING ONTOGENESIS?

PETER H. WOLFF

Thelen and Ulrich's *Monograph* will be of great interest at several levels for students of human behavioral development and motor coordination. For example, the empirical investigations of stepping movements represent a major contribution to our understanding of the dynamics of motor coordination. Similarly, the theoretical sections introduce readers to a radically different perspective on behavioral development. At the same time, they address an enduring and still unresolved problem inherent to all the developmental sciences. How do organisms induce qualitatively new behavioral forms during ontogenesis whose novel structures and functional properties cannot be deduced from or reduced to the characteristics of the organism's antecedent conditions? By challenging cherished theories of psychological and behavioral development, this section motivates us to reexamine critically many of the a priori theoretical premises generally taken for granted in developmental research. The investigators' reasons for adopting a "dynamic perspective" research strategy and for introducing the somewhat idiosyncratic methods of data collection, analysis, and interpretation that are inherent to this perspective emerge most clearly when the sections are studied in relation to one another. However, for purposes of exposition, it may be useful to examine the various levels of description separately.

The empirical sections describe a series of elegant experiments on the precursors of independent walking that invite direct comparison with the classic studies of neuromotor maturation by Gesell (1954), McGraw (1966), and others. At the same time, they highlight differences between traditional studies of motor development and the dynamic research strategy. The study sample is limited to nine healthy infants examined repeatedly over the first year of life; no age means and standard deviations are reported, and the significance of changes over time is not tested by conventional statistical procedures. The investigators persuasively argue that "variability is as much a part of the developmental story as is uniformity" (p. 45). Because conventional statistical smoothing procedures tend to obscure essential information about the *process*, in contrast to the products, of development, they have instead plotted the developmental trajectories of stepping movements separately for each infant and explored what general principles of behavioral coordination might best explain the fact that, despite variable rates, all healthy children will eventually reach the same stable end point of a flexible bipedal gait that can adapt itself to unpredictable variations of environmental requirements. In this sense, their strategy resembles Piaget's methods of inferring the general course of early sensorimotor intelligence from observational data on three children who exhibited different rates and alternative pathways to achieve the same strategies for cognizing the world. Had the authors applied a conventional strategy of computing group statistics to determine the mean age when most infants reach any particular level of performance, they might have concluded that the species-typical development of stepping movements and bipedal gait is controlled by a priori maturational time tables. By examining the developmental trajectories of individual infants under a dynamic systems research strategy, they are led to the very different conclusion that coordinated motor patterns exhibit a propensity for self-organization and self-correction in the face of variable and unpredictable environmental demands and that the invariance of developmental transformation may reside in principles of spontaneous pattern formation rather than in specific a priori prescriptions.

Many previous investigators have recorded and described the ubiquity of individual differences in almost all dimensions of human behavioral development. However, the authors of this *Monograph* have taken a critical next step by not only identifying the range of individual variations but at the same time clarifying the presumably universal processes of pattern formation and motor coordination that can encompass the origins of individual differences in developmental rate. In other words, they do not leave us with the theoretically empty truism that infants exhibit individual differences in many behavioral domains.

In contrast to "ethologically informed" studies of human behavioral development, in which experimental manipulations are introduced only after the species typicality of naturally occurring motor behavior under free-field conditions has been described, Thelen and Ulrich deliberately use perturbation experiments at each point in their longitudinal study. For example, the treadmill device was designed to dislodge naturally occurring stepping movements from their conservative organization in order to define the regions of dynamic instability in coordinated behavior where minimal quantitative changes in one or more relevant control parameters would precipitate a phase transition and in order to analyze the dynamic characteristics of the emerging pattern of coordination once the critical region of instability has been exceeded. In this way, the investigators were able to characterize the intrinsic dynamics of behavioral coordination at successive points during development and to plot developmental changes in these dynamics. The perturbation experiments were therefore the essential means for uncovering the "hidden precursors" of bipedal walking that would not be evident to direct observation under free-field conditions. Similarly, by varying the speed of the treadmill belt and changing the speed ratios of the split belts that activated the right and left feet, the investigators were able to test experimentally which control parameters induced the phase transitions in bipedal stepping movements at different points in development without themselves determining what novel patterns of behavioral coordination would be realized.

To analyze their findings from a dynamic perspective, the investigators introduced "contentless" units of description, based on concepts such as point attractors, limit cycles, phase transition, and collective variables, that were largely imported from the mathematical description of complex physical systems. At first glance, the use of such unfamiliar terms may seem like a grandiose redescription of mundane behavioral phenomena, but the sections on data analysis and the interpretation of findings indicate why such concepts were essential to analyze the dynamics of motor coordination and to identify critical regions of instability where minimal changes in one or more control parameters would precipitate major phase transitions and qualitative shifts to a novel pattern of motor coordination. The theoretical section that proposes to "introduce a new developmental theory and a strategy for examining developmental processes explicitly derived from the contemporary study of dynamic systems," that is, *synergetics* or *dynamic pattern theory*" (p. 1), provides a more detailed formal justification for using such abstract units of description.

Biology and developmental psychology have long recognized that the behavior of organisms cannot be reduced to a few fundamental interactions in terms of linear deterministic laws. However, until recently, biologists lacked the formal analytic tools to investigate the complexity of behavioral forms and their developmental transformations experimentally and therefore resorted to metaphors or undefined constructs in order to "explain" the essential differences between man-made machines and living things. Only recently have biologists systematically applied nonlinear differential equations to describe, analyze, and experimentally predict the behavior of complex organisms. With the help of powerful new conceptual tools, biologists and behavioral scientists can now describe the finer details of complexity in selected naturally occurring phenomena that previously had been ignored because they did not fit a mode of biological systems as deterministic machines. Further, they can now carry out theoretically informed experiments to elucidate phenomena of self-organization, spontaneous pattern formation, and the induction of novel behavioral forms and functions without recourse to explanations in terms of a priori genetic, neurological, or environmental prescriptions to account for such phenomena.

Thelen and Ulrich have taken a major step forward by introducing the same dynamic systems perspective to the study of behavioral coordination in infants. Their findings challenge many established theories of human behavioral development, including traditional neurological-maturationist and behaviorist worldviews as well as the currently accepted epigenetic-constructivist, cognitive, and information-processing models. The findings further suggest that the investigation of behavioral development need no longer be constrained by theoretical formulations that ultimately fall back on a priori and essentially untestable notions of genetic programs, maturational timetables, hard-wired central pattern generators, hard-programmed brain modules, or environmental constancies (schedules of reinforcement) to account for the invariance of normal behavioral development under qualitatively different initial conditions.

However, applications of the dynamic perspective to concrete developmental questions entail a great deal more than theoretical arm waving or the metaphoric redescription of familiar phenomena in an obscure language. As the methods of data analysis reported in the manuscript demonstrate, any meaningful application of the dynamic systems research strategy depends critically on theoretically informed experiments and their meticulously detailed qualitative and quantitative analyses by formal procedures that differ in principle from the frequency statistics conventionally applied to developmental data.

The dynamic systems perspective is presented here as a new developmental theory with potential application to many domains of development that had until now remained opaque to conventional research strategies. However, the authors recognize that the generality of its scope may be symptomatic of the perspective's inherent weakness. It is relatively uninformative, for example, about the finer details of what actually occurs in any specific behavioral domain during development, and it can make only very limited a priori predictions about the particular behavioral transformations that will be realized after the transition from one marginally stable state to another. Furthermore, as the authors note, the perspective is currently in a much less developed state of elaboration than established approaches to the study of development, and the boundaries of its general relevance for understanding human behavioral development therefore remain to be explored. Even within the domain of motor coordination, for example, many behavioral patterns that undergo radical developmental transformations may simply not lend themselves to rigorous formal analysis. The same caution applies more emphatically to investigations of the development of communicative language, symbolic thought, operational logic, and social-emotional development. On the other hand, one should not dismiss outright the possibility that component aspects of even these specialized topics can in principle benefit from a critical reexamination of their basic theoretical premises and a reanalysis of the important findings from a dynamic systems perspective. The proof of the pudding evidently remains in the eating.

In sum, the dynamic systems perspective is not a theory of everything and should not be construed as such. As currently articulated, it may not even be "a new developmental theory" in any conventional sense; instead, it may comprise a set of formal procedural guidelines for investigating the inherent complexity of a wide range of naturally occurring phenomena. Nevertheless, the *Monograph* leaves one with the distinct impression that a rigorous application of the proposed research strategy with an appropriate respect for the complexity of dynamic systems can lead us to radically different methods of developmental investigation and radically different theoretical perspectives on the processes of novel pattern formation. By this route, the perspective opens the way for experimentally investigating phenomena that were previously inaccessible because of our theoretical preconceptions.

References

Gesell, A. (1954). The ontogenesis of infant behavior. In L. Carmichael (Ed.), *Manual of child psychology* (2d ed.). New York: Wiley.
McGraw, M. B. (1966). *The neuromuscular maturation of the human infant.* New York: Hafner.

CONTRIBUTORS

Esther Thelen (Ph.D. 1977, University of Missouri) is professor of psychology at Indiana University. Her research interests have focused on patterned movement in infants and on the acquisition of early motor skills. She is currently writing a book (with Linda B. Smith) applying dynamic systems theory to the development of cognition and action.

Beverly D. Ulrich (Ph.D. 1984, Michigan State University) is assistant professor of kinesiology at Indiana University. Her work on the development of motor skills focuses on two main interests: dynamic systems theory as an approach to understanding the development of nonhandicapped and Down syndrome infants and the interface between theory and intervention to reduce developmental delay.

Peter H. Wolff (M.D., 1950, University of Chicago) is professor of psychiatry at the Harvard Medical School. His research interests include the development of behavioral coordination and behavioral states in infancy and the temporal organization of motor behavior in infants, children, and adults.

STATEMENT OF EDITORIAL POLICY

The *Monographs* series is intended as an outlet for major reports of developmental research that generate authoritative new findings and use these to foster a fresh and/or better-integrated perspective on some conceptually significant issue or controversy. Submissions from programmatic research projects are particularly welcome; these may consist of individually or group-authored reports of findings from some single large-scale investigation or of a sequence of experiments centering on some particular question. Multiauthored sets of independent studies that center on the same underlying question can also be appropriate; a critical requirement in such instances is that the various authors address common issues and that the contribution arising from the set as a whole be both unique and substantial. In essence, irrespective of how it may be framed, any work that contributes significant data and/or extends developmental thinking will be taken under editorial consideration.

Submissions should contain a minimum of 80 manuscript pages (including tables and references); the upper limit of 150–175 pages is much more flexible (please submit four copies; a copy of every submission and associated correspondence is deposited eventually in the archives of the SRCD). Neither membership in the Society for Research in Child Development nor affiliation with the academic discipline of psychology are relevant; the significance of the work in extending developmental theory and in contributing new empirical information is by far the most crucial consideration. Because the aim of the series is not only to advance knowledge on specialized topics but also to enhance cross-fertilization among disciplines or subfields, it is important that the links between the specific issues under study and larger questions relating to developmental processes emerge as clearly to the general reader as to specialists on the given topic.

Potential authors who may be unsure whether the manuscript they are planning would make an appropriate submission are invited to draft an outline of what they propose and send it to the Editor for assessment.

This mechanism, as well as a more detailed description of all editorial policies, evaluation processes, and format requirements, is given in the "Guidelines for the Preparation of *Monographs* Submissions," which can be obtained by writing to Wanda C. Bronson, Institute of Human Development, 1203 Tolman Hall, University of California, Berkeley, CA 94720.